Artificial Intelligence Applications in a Pandemic

Artificial Intelligence Applications in a Pandemic
COVID-19

Edited by
Salah-ddine Krit
Vrijendra Singh
Mohamed Elhoseny
Yashbir Singh

CRC Press
Taylor & Francis Group
Boca Raton London New York

CRC Press is an imprint of the
Taylor & Francis Group, an informa business

First edition published 2022
by CRC Press
6000 Broken Sound Parkway NW, Suite 300, Boca Raton, FL 33487-2742

and by CRC Press
2 Park Square, Milton Park, Abingdon, Oxon, OX14 4RN

© 2022 selection and editorial matter, Salah-ddine Krit, Vrijendra Singh, Mohamed Elhoseny, Yashbir Singh; individual chapters, the contributors.

CRC Press is an imprint of Taylor & Francis Group, LLC

Library of Congress Cataloging-in-Publication Data

ISBN: 978-0-367-64449-9 (hbk)
ISBN: 978-0-367-64774-2 (pbk)
ISBN: 978-1-003-12621-8 (ebk)

DOI: 10.1201/9781003126218

Typeset in Palatino LT Std
by Newgen Publishing UK

Contents

Preface

When we began to embark on the path that eventually led to this book, and while trying to provide an accurate and possibly interesting way to explain the role of AI in this pandemic, it seemed that saying "Artificial Intelligence (AI) Applications in a Pandemic: COVID-19" could be a good way to start.

To be able to study and learn about the application of artificial intelligence in this COVID-19 has certainly been an honor and a great challenge for everybody.

Hopefully, in this book, we will be able to fulfil the main objective and demonstrate the unique and useful applications of AI, in particular to COVID-19, and to discuss new ideas and the importance of the artificial intelligence in this "COVID-19 era," expecting to examine the challenge of advancement of artificial intelligence in medical science. This book will be very effective to undergraduate and postgraduate studies, to analysts and academicians, and for high-quality researches of digital healthcare technologies.

My role in this book has been to collaborate with many good researchers around the world to bring their struggles, challenges, and philosophies before the readers. I am deeply grateful to them all for taking the time from their busy lives to write these case studies and work with me through the long editorial process. I also extend my gratitude to all my colleagues in the domains of artificial intelligence to help resolve the problems of the COVID-19 era.

Editors

Salah-ddine Krit is an associate professor at the Polydisciplinary Faculty of Ouarzazate, Ibn Zohr University, Agadir, Morocco. Dr. Krit is director of the Engineering Science and Energies Laboratory and head of Department of Mathematics, Informatics, and Management. He received PhD degrees in software engineering from Sidi Mohammed Ben Abdellah university, Fez, Morocco, in 2004 and 2009, respectively. During 2002–2008, he worked as an engineer team leader in audio and power management in Integrated Circuits (ICs) research, design, simulation, and layout of analog and digital blocks dedicated to mobile phone and satellite communication systems using Cadence, Eldo, Orcad, and VHDL-AMS technology. Dr. Krit has authored or co-authored over 130 journal articles, conference proceedings and book chapters published by IEEE, Elsevier, Springer, Taylor & Francis, IGI Global, and Inderscience. His research interests include Wireless Sensor Networks, Network Security, Smart-Homes, Smart-Cities, the Internet of Things, Business Intelligence, Big Data, Digital Money, Microelectronics, and Renewable Energies.

Vrijendra Singhi is an associate professor and head of the Department of Information Technology, Indian Institute of Information Technology Allahabad, Deoghat, Jhalwa. He is an active member of IEEE and the International Association of Engineers. He has more than 351 research citation indices with Google Scholar and has authored various research papers in reputed conferences and journals, including *Web of Science* and *Scopus*. He is a reviewer for many international journals and a member of organizing and scientific committees for many international conferences.

Mohamed Elhoseny is an associate professor at the University of Sharjah, UAE. Elhoseny is an ACM Distinguished Speaker and IEEE Senior Member. His research interests include Smart Cities, Network Security, Artificial Intelligence, the Internet of Things, and Intelligent Systems. Elhoseny is the founder and the Editor-in-Chief of IJSSTA journal published by IGI Global. Also, he is an associate editor at several Q1 journals such as *IEEE Journal of Biomedical and Health Informatics, IEEE Access, Scientific Reports, IEEE Future Directions, Remote Sensing, International Journal of E-services and Mobile Applications, and Human-centric Computing and Information Sciences.* Moreover, he served as the co-chair, the publication chair, the program chair, and a track chair for several international conferences published by recognized publishers such as IEEE and Springer. Elhoseny is the Editor-in-Chief of The Sensors Communication for Urban Intelligence, CRC Press-Taylor & Francis book series, and the Editor-in-Chief of the Distributed Sensing and Intelligent Systems, CRC Press-Taylor & Francis book series.

Yashbir Singh is a research fellow in the Department of Radiology at the Mayo Clinic, Rochester, Minnesota. He previously was a medical scientist at the Heart and Vascular Institute, West Virginia University, involved with the cardiac imaging strain data using topological data analysis. He completed his PhD in biomedical engineering in Taiwan, with an interest in contributing to a deeper understanding of cardiac malfunction using computational techniques. His PhD thesis was involved with scar tissue identification and creation of the scar geometry using topological data analysis and deep learning approach with advanced image analysis. During his PhD, he received a Deutscher Akademischer Austauschdienst, German Academic Exchange Service DAAD fellowship at Magdeburg University, Germany, with an interest in non-invasive sensors and signal processing for guiding medical interventional devices.

1

Role of Artificial Intelligence in COVID-19

S. Lalitha, H. T. Bhavana, K. N. Madhusudhan, Prascheth, and Harshitha

CONTENTS

1.1 Introduction

Coronavirus disease (COVID-19) is an inflammation infection originated by novel coronavirus. Cold, cough, fever, and loss of smell and taste are general symptoms of COVID-19 and, in critical cases, respiratory problems. It has been declared a worldwide pandemic. The important phases of human life, such as transportation, education, health sector, marketing have been changed due to the attack of COVID-19. Those infected with COVID-19 suffer from respiratory problems but can recovered with proper medical aid. The virus can easily transmit from human to human, hence, it is one of the most dangerous viruses among all virus families. A total of 214 countries have been suffering from this deadly disease. As of September 10, 2020, the worldwide confirmed coronavirus cases numbered 27.5 million, and 894,983 deaths have occurred. The United States had the record for the most cases of COVID-19,

DOI: 10.1201/9781003126218-1

nearly 6.3 million confirmed, and 190,856 deaths. Italy, Brazil, India, Spain, and Russia are some other countries that are enormously affected.

Artificial Intelligence (AI) is a trending technology for a large number of applications in several fields. Some major areas where AI can be utilized are medical diagnosis and healthcare, image processing, natural language processing, banking sector, cyber security systems, and so forth. AI has number of branches based on different approaches of solving problems, among them Machine Learning (ML) and Deep Learning (DL) are the main categories. ML extracts meaningful features from given data and, hence, requires huge amount of data input. DL algorithms are able to solve complex problems from simple representations. DL uses multiple layers to learn data in deep manner. AI can definitely help us in addressing the problems raised by COVID-19 pandemic [1].

1.2   Background

COVID-19 is caused by severe acute respiratory syndrome, coronavirus 2 (SARS-CoV-2), a beta coronavirus. Wuhan city in China was the first city where this novel coronavirus reported on 31 December, 2019. Now almost all countries are suffering from the spread of COVID-19. The number of infected cases are increasing day by day, and no sign of reduction in infected and death cases. Since the numbers of new cases and fatality rate are very large, the situation is not under control. The World Health Organization (WHO) is applying more effort in analysing and assessing the risk to the highest level.

Government have put in so much effort for a solution to fight the COVID-19 pandemic, as fully developed countries also facing the worst scenario. Governments' attempts are mainly to stop the spreading of the virus. Hence, measures have been adopted by government, such as lock down of a particular state or country, the sealing of specific regions, and economic support packages to give financial aid in this pandemic situation, strengthen the health care system to face the most terrible consequences due to COVID-19, provide adaptive policies to citizens, and so forth. People are also advised to stay safe and healthy, avoid public gatherings, use face masks, wash/sanitize hands frequently, follow social distancing customs, and report health issues/symptoms to regional health centers [3].

Infected patients must be screened effectively, and providing immediate medical aid is very much essential to fight the spreading of COVID-19. The clinical screening method for COVID-19 patients is reverse transcription polymerase chain reaction (RT-PCR), which uses respiratory samples for testing. This technique is manual and complicated. Because of shortage of supply of RT-PCR there is delay in the prevention of disease. This delay then leads to healthy people communicating with infected patients and spreading the infection further. The cost of RT-PCR is also high, which makes the international scenario still worse [4].

Recent advances in artificial intelligence show that many biomedical health problems and hurdles can be solved by applying ML and DL algorithms. Significant image features can be revealed by AI, features that are not evident in the original images. Powerful algorithms of neural networks have been proven tremendously beneficial in extracting the features and learning. It is extensively included in the research community as many communicable diseases such as COVID-19, SARS, HIV, EBOLA, and non-communicable diseases such as diabetes, cancer, stroke, and so forth, can be diagnosed by applying powerful AI algorithms. The availability of a large amount of healthcare data can provide the useful features for AI algorithms to learn and to analyse the new problems and the handling of risky situations. Some of the repetitive jobs can be carried out by AI, such as analyzing medical tests, X-Ray images, CT scans, or available data that can reduce the human error in clinical diagnosis. Intelligent AI algorithms can provide up-to-date medical information from medical journals, and regular clinical procedures and textbooks to physicians. Drug prediction and selection for particular diseases based on the clinical symptoms can be done by AI [8].

1.3 Algorithms Used in Artificial Intelligence

AI has a diversity of applications in clinical studies and healthcare services. All healthcare applications under AI can be generally classified into two main categories [8].

1. Structured data analysis, which includes biomarkers, images, and genes.
2. Unstructured data analysis, which comprises medical journals, survey reports, and notes.

Structured data analysis is based on machine learning and deep learning algorithms. The unstructured data analysis relies on Natural Language Processing algorithms. Figure 1.1 shows the ML and DL processing in healthcare systems. Machine learning algorithms usually extract features from available data known as patients' "traits," and is the outcome of patients' diagnosis. ML has different algorithms that serve different purposes [8]. Some of the algorithms and techniques used in development of machine learning models for COVID-19:

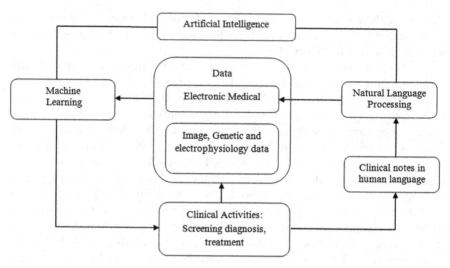

FIGURE 1.1
Classification of AI in Healthcare system.

- Support Vector Machine
- Neural Networks
- Decision trees
- KNN (K Nearest Neighbours)
- Logistic regression
- Random forest
- Linear regression

1.3.1   Support Vector Machine (SVM)

SVM is primarily used in classification problems that need data sets to be divided into two classes based on a separating line, or a "hyperplane." It takes data as input and as output gives a separating line that separates those two classes. The best solution is to provide the hyperplane with the greatest possible margin so that new data can be classified correctly. The points closest to the data plane are known as support vectors, the change in position of which would alter the position of the hyperplane itself [9]. Figure 1.2 shows the flow chart of SVM. The hyperplane will be defined based on the input data set and new data will be classified based on the separating line. Figure 1.3 illustrates a simple example of classifying the squares and ovals based on hyperplane. Here, the best hyperplane would be the right-tilted line rather than the vertical line as it provides maximum margin so that squares and ovals can be classified correctly.

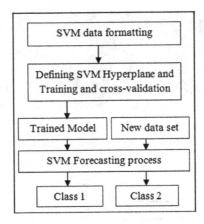

FIGURE 1.2
Flowchart of SVM.

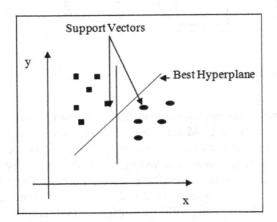

FIGURE 1.3
Optimal hyperplane using SVM algorithm.

SVM in healthcare systems are mainly used for protein classification, image segmentation, and text categorization.

Case study: A classification method has been designed using the X-ray images of COVID-19, healthy people versus pneumonia to screen out healthy and pneumonia effected people [10]. The dataset contains 127 validated COVID-19 images, 127 pneumonia (63 bacterial pneumonia and 64 viral pneumonia) images and 127 healthy images. Finally, only 48 COVID-19 frontal view X-ray images were considered, which includes posteroanterior (PA) and anteroposterior (AP) views. Lung diseases can be examined with the help of views of these images. Figure 1.4 represents X-ray images of COVID-19, bacterial pneumonia, viral pneumonia, and a normal person [10].

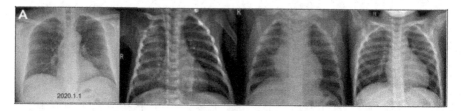

FIGURE 1.4
X-ray images of (a) COVID-19, (b) bacterial pneumonia, (c) viral pneumonia, (d) normal.

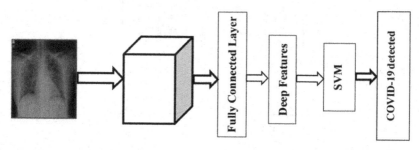

FIGURE 1.5
Detection of COVID-19 using SVM.

Pre-trained networks are used to extract the features using dataset. These deep features are fed to SVM for classification; finally, image-classification methods based on a traditional approach has been carried out for detection of COVID-19. The fully connected layer provides the deep features and will be fed to a classifier for training. These deep features will be used by SVM to categorize the input into different classes. Figure 1.5. shows the method of detection of COVID-19, using SVM by applying deep features.

The author has compared the performance of 13 classifier models, among them ResNet plus SVM is better and could be able to provide minimum and maximum accuracy of 92 percent and 98.66 percent respectively.

1.3.2 Convoluted Neural Networks (CNN)

Deep learning is considered as a division of the broader family of machine learning. The most commonly used architectures are deep neural networks. Under deep neural networks, Convolutional Neural Networks (CNN) are widely popular. CNN is mainly used on images. They learn meaningful features from images by assigning weights to various aspects of the image [11].

1. The important block in CNN is convolution operation; it is a mathematical operation that combines two sets of information. Convolution

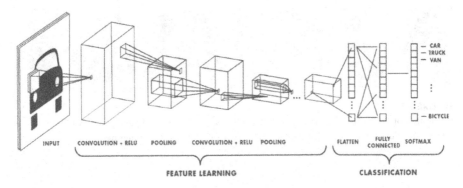

FIGURE 1.6
General steps involved in CNN.

is applied by sliding appropriate filters all over the input image to produce a feature map.

2. The second step makes use of Rectified Linear Function (ReLu) activation function. This increases nonlinearity in the images and, hence, it progresses the color.

3. The pooling step reduces the dimensionality, which leads to reduction in the number of parameters and, hence, to less training time and reduces over fitting.

4. The two-dimensional image matrix will be converted into a linear vector during the flattening process.

5. The final step is forming a fully connected layer. It will be packed completely with output of the previous layer and forms the desired number of Figure 1.6 provides the general steps involved in CNN which categorizes the vehicles into car, truck, van, and so forth.

Case study: Computed tomography (CT) can help in diagnosing COVID-19 patients. This work classifies and compares COVID and other viral pneumonia patients, using Convolutional Neural Networks [12]. Chest-high resolution CT (CHCT) examination was done on all patients. The ground glassy opacity (GGO), "crazy-paving" pattern with a peripheral distribution and multifocal patchy amalgamation are the main chest CT findings of COVID-19. Figure 1.7 shows chest Computer Tomographic images of COVID-19, H1N1 and viral pneumonia patients. The red arrows indicates GGO at different lobes in lungs [12].

In this work, CT images of the region of infection were initially defined per slice by radiologists for further analysis; 1020 CT slices were considered in this work. These slices were extracted from 108 subjects with confirmed COVID-19 cases and 86 subjects with unusual pneumonia. Ten popular neural networks were applied to classify COVID and non-COVID patients.

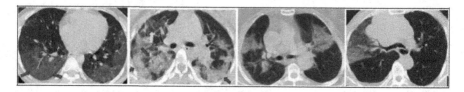

FIGURE 1.7

CT images of (a) 28-year-old patient with COVID-19, (b) 33-year-old patient with COVID-19, (c) 81-year-old H1N1 patient influenza, (d) 72-year-old patient with pneumonia.

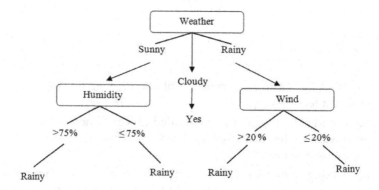

FIGURE 1.8

Decision tree for weather prediction.

ResNet-101 provided the highest sensitivity and accuracy of 99.51 percent and proved to be promising CNN for diagnosis of COVID. The author suggests using this method as adjuvant during CT imaging in the radiology department.

1.3.3 Decision Trees

Tree is a very powerful data structure for classification. Decision tree algorithms work for both continuous and categorical variables. It is a tree-like graph where nodes represent a place where an attribute will be selected and a question will be asked. Leaves of the tree represent the class label, which will be actual output. The given example will be classified by sorting them down the tree from root to leaf node. Each node will be a test case for some attribute [13].

The child cases will get formed at each node after running a classification algorithm. At the beginning the entire training set is considered as a root. As the algorithm runs, they are classified into various leaves of the tree. Figure 1.8 shows a simple illustration of the decision tree in predicting weather. In COVID 19 all the parameters are interlinked in a complex way. Using decision tree algorithms the entire population can be split into groups and linear and nonlinear relationships can be mapped.

Case study: The Chest Radiography X-ray (CXR) can be used as an alternate diagnosis method for COVID-19 related pneumonia using a decision tree classifier from CXR images [14]. The author has considered normal and abnormal X-ray images and corresponding data of different types of lung diseases. These datasets are provided by different health organizations based on pathological diagnosis reviewed by radiologists. This work contains the binary decision trees. Every tree structure is trained by deep learning models. This classifies the images based on four-level decision trees. Four augmented X-ray imaging radiography systems (AXIRs) are used. The CXR images will be classified as normal or abnormal in the first decision tree. The second tree classifies the images containing the symptoms of tuberculosis from the abnormal images. The third tree identifies the symptoms of COVID 19. The accuracy of first, second, and third decision trees are 98 percent, 80 percent and 95 percent respectively. This method can be applied before the conventional testing method of COVID-19 as pre-screening process. Figure 1.9 shows the decision tree for classifying COVID and non-COVID patients.

1.3.4   K-Nearest Neighbours (KNN)

KNN is a widely used classifier algorithm. First the model is trained on the dataset, and the "K" value is chosen for predicting new test cases. Prediction works by calculating the distance function from a known point (labelled "input data") plotted while training. The algorithm first finds "K" neighbours to a point that is at the least distance (Hamming distance, Manhattan distance, Euclidean distance, etc.). New data is mapped to a classification set based on the class of maximum neighbours. It is a computationally expensive algorithm and needs data to be pre-processed before prediction. Figure 1.10 depicts the general conception of K-nearest neighbour, The new example to classify will be added to either Class A or Class B based on three steps [15].

1. Distance calculation
2. Finding Closest Neighbours
3. Vote for labels

Case-study: KNN classifier is more flexible and can be applied for complex classification problems. The author in [16] has introduced a model known as COVID-19 Patient Detection Strategy (CPDS). Initially the features from CT images are extracted through Gray Level Co-occurrence Matrix (GLCM), and then the effective features are selected from Hybrid Feature Selection Methodology (HFSM). Finally based on the selected features fast and accurate detection of COVID-19 is done through Enhanced KNN classifier (EKNN).

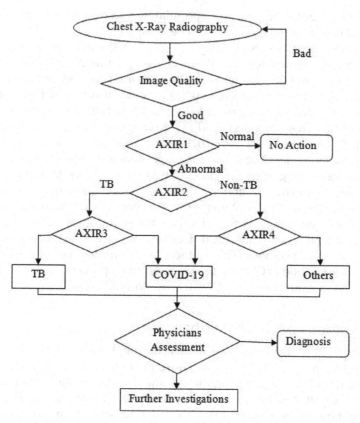

FIGURE 1.9
Flow chart of analysis of COVID and Non-COVID patients using decision tree.

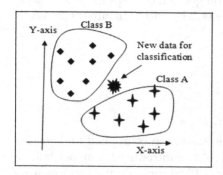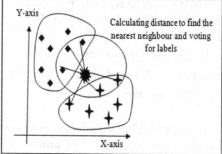

FIGURE 1.10
K-nearest Neighbours.

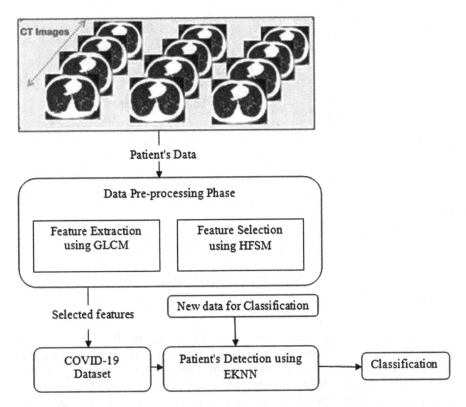

FIGURE 1.11
COVID-19 detection using CPDS strategy.

This algorithm can achieve 96 percent accuracy and also provide fast and more accurate results. Figure 1.11 shows the steps involved in detection of a COVID-19 patient based on CPDS method.

1.3.5 Logistic Regression

Logistic regression is a type of classification algorithm which, after studying the patterns in some independent variables, answers a yes/no question. It transforms the output using logistic sigmoid function and gives a probability value that can then be mapped to two or more discrete classes. This sigmoid function maps any real value into another value between 0 and 1. It gives the probability of an event happening and is extensively used in the medical field. Figure 1.12 shows the binary sigmoid function of variable x. The horizontal line indicates decision boundary [17]. Binary sigmoid function can have two values, 0 and 1. Classes can be categorized based on the sigmoid function as

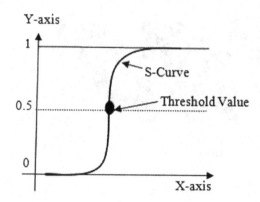

FIGURE 1.12
Logistic regression.

Sigmoid(x) ≥ 0.5, class1

Sigmoid(x) ≤ 0.5, class 2

After studying various parameters available in the medical history of a person, it can tell the probability of a person's contracting a disease. We have seen patterns such as mortality rates in different age groups, asymptomatic patient rates, and COVID severely affecting patients suffering from long-term diseases. By this algorithm we can predict people who are at higher risk and offer them better medical resources

Whether COVID-19 patients at high risk will recover or not depends upon how early they are treated. It was reported that the fatality rate of critical cases was about 61.5 percent, and the risk is highest with increasing age and with certain health conditions. A large number of patients have been confirmed worldwide who are without effective treatment, and the medical systems are under great pressure with severe shortages of intensive care units and other resources.

Case-study: A logistic regression model has been designed to forecast the death rate of COVID-19 patients [18]. This model uses three important blood biomarkers, including High-Sensitivity C-reactive protein, Lymphocyte (%), and Lactic Dehydrogenase, and their interactions. This algorithm is accurate and explainable, which predicts the probability of a fatality as between zero and one, rather than a binary. When the fatality probability was found to be 0.8, the model was able to predict the subject's fatality by more than 11.3 days on average, with maximally 34.91 days in advance. The accuracy score obtained was 93.92 percent. This type of model helps to identify the patients with highest risk of fatality and, hence, aid the existing medical systems across the globe to plan use of resources accordingly.

1.3.6 Linear Regression

Regression is a process of modelling a target value based on independent predictors. This algorithm is basically used in finding the relationship between the variables and, hence, prediction of a future value will be done. Linear regression is a simple yet powerful algorithm that is trained with continuous data. This algorithm shows a linear relationship between two variables. Among these two variables, one will be dependent (Y) on another, which will be the independent (X) variable. This algorithm finds how the value of the dependent variable is changing according to the value of the independent variable [19].

Figure 13 shows the linear regression graph, where the linear relationship between two variables are shown. Mathematically, linear regression can be represented as Y = aX+b, where X,Y are the independent and dependent variables. The slope of the line is b, and a is the intercept (the value of y when x=0). These coefficients, a and b, are derived by reducing the sum of squared difference of the distance between the regression line and data points.

Linear regression plays a significant role in statistics. Since no country has yet invented any medicine to cure the disease completely, management of this epidemic is to reduce its peak, also known as levelling the epidemic curve [19]. The linear regression model can be applied in various aspects of treating and managing COVID-19.

Case study: The linear regression model was used to trace the drift linked to death counts expected at the 5th and 6th week of COVID-19. This also determines the rate of spread of COVID-19 disease in next 7 days for better management by doctors and various government organizations [20]. A validated database was used to get the Indian and global data related to the coronavirus. The author has considered developing a database of COVID-19 from March 1 to April 11, 2020, and predicted next one for number of patients suffering from rising COVID. Six linear regression analysis models, namely quadratic, third degree, fourth degree, fifth degree, and sixth degree were used along with auto-regression techniques to improve the predictive

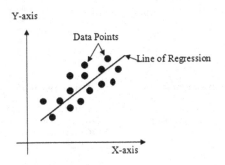

FIGURE 1.13
Linear regression.

ability of the regression model. In these six models the six-degree polynomial is good, as the root mean square error is very less. Hence, it is used for forecasting the next 6 days' scenario for COVID-19 data analysis in India.

1.4   Applications of AI for Fighting Covid-19

Artificial intelligence refers to a field of computer science devoted to a system that is designed to carry out tasks that require intelligence and has aided in many areas of COVID-19. AI can help in identifying the coronavirus in early stages and also in monitoring the medical condition of the infected patient. Analyzing the available data research on this virus can be facilitated [20]. The latent contributions of AI during COVID-19 should be known in order to acquire a good understanding of this technology, some aspects of which are discussed below in Figure 1.14.

1.4.1   Issues and Challenges in AI COVID-19

Healthcare Industry Issues: Conventional doctors may oppose the new innovations, and also patients may retrospect the AI-Based Decision making.

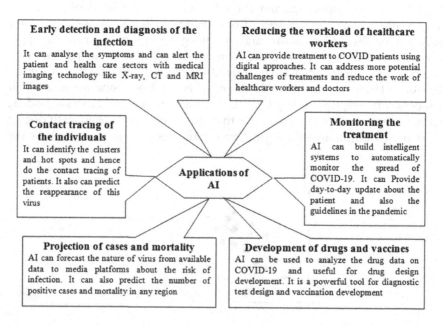

FIGURE 1.14
Applications of AI in managing COVID-19.

A lot of effort would be needed to instill confidence in patients and doctors who are not ready to upgrade.

Technology-related Issues: Since the virus mutates, the algorithms built may not be reliable on the quality of data in decision making and processing. Also, maintaining a huge data base and managing it is a challenge.

Socio-cultural Issues in Technology Implementation: Though a large part of the population in India is accustomed in using the internet and mobile technology, there is still an ignorance towards the change in information technology, as people think it would ruin their traditions and customs. An example is a gender-based mobile technology usage. Predominantly, men will have the upper hand in using the mobiles, retarding women using them as it would disturb the socio-cultural harmony.

Regulatory and Ethical issues: Information security and privacy are the main constraints with the increase in the usage of devices–wearables, which are prone to hacking. Also, AI might change the conventional relationship between the doctor and the patient.

1.5   Conclusion

Healthcare research institutes and firms are racing towards finding a unique and effective technique to get a grip on this deadly virus, and to assist them in adequate and reliable inputs in real-time to break the binge chain. AI operates in a capable fashion to almost simulate human intelligence. It would show an important and significant role in understanding, and proposing the improvement of, a vaccine for the deadly COVID-19 virus. This result-oriented technology is also used for proper identifying, screening, evaluating, predicting, and tracing of suspected and actual future subjects. The important applications are also implemented to track data of all kinds of cases, including confirmed, recovered, and death cases.

References

[1] Quoc-Viet Pham, Dinh C. Nguyen, Thien Huynh-The, Won-Joo Hwang, and Pubudu N. Pathirana. "Artificial Intelligence (AI) and Big Data for Coronavirus (COVID-19)".

[2] "Pandemic: A survey on the state-of-the-arts", IEEE access, DOI:10.1109/ ACCESS.2020.3009328.

[3] www.who.int/emergencies/diseases/novel-coronavirus-2019.

[4] R. Sujath, Jyotir Moy Chatterjee, and Aboul Ella Hassanien, "A machine learning forecasting model for COVID-19 pandemic in India", https://doi.org/10.1007/s00477-020-01827-8.

[5] Muhammad E. H. Chowdhury, Tawsifur Rahman, Amith Khandakar, Rashid Mazhar, Muhammad Abdul Kadir, Zaid Bin Mahbub, Khandakar R. Islam, Muhammad Salman Khan, Atif Iqbal, Nasser Al-Emad, Mamun Bin Ibne Reaz, M. T. Islam, "Can AI help in screening Viral and COVID-19 pneumonia?", DOI:10.1109/ACCESS.2020.3010287, IEEE Access.

[6] Samuel Lalmuanawma, Jamal Hussain, and Lalrinfela Chhakchhuak, "Applications of machine learning and artificial intelligence for Covid-19 (SARS CoV-2) pandemic: A review", https://doi.org/10.1016/j.chaos.2020.110059.

[7] https://medium.com/sciforce/top-ai-algorithms-for-healthcare-aa500 7ffa330.

[8] https://towardsdatascience.com/https-medium-com-pupalerushikesh-svm-f4b42800e989.

[9] Prabira Kumar Sethy, Santi Kumari Behera, Pradyumna Kumar Ratha, and Preesat Biswas, "Detection of coronavirus disease (COVID-19) based on Deep Features and Support Vector Machine", doi.org/10.33889/IJMEMS.2020.5.4.052.

[10] https://towardsdatascience.com/basics-of-the-classic-cnn-a3dce1225add.

[11] Singh, Y., Shakyawar, D., & Hu, W. (2020). "An automated method for detecting the scar tissue in the left ventricular endocardial wall using deep learning approach." *Current Medical Imaging*, 16(3), 206–213.

[12] www.geeksforgeeks.org/decision-tree/.

[13] Seung Hoon Yoo, Hui Geng, Tin Lok Chiu, Siu Ki Yu, Dae Chul Cho, Jin Heo, Min Sung Choi, Il Hyun Choi, Cong Cung Van, Nguen V. Nhung, Byung Jun Min, and Ho Lee, "Deep Learning-Based Decision-Tree Classifier for COVID-19 diagnosis from chest X-ray imaging", doi:10.3389/fmed.2020.00427.

[14] https://medium.com/machine-learning-researcher/k-nearest-neighbors-in-machine-learning-e794014abd2a.

[15] Warda M. Shaban, Asmaa H. Rabie, Ahmed I. Saleh, M. A. Abo-Elsoud, "A new COVID-19 Patients Detection Strategy (CPDS) based on hybrid feature selection and enhanced KNN classifier", doi.org/10.1016/j.knosys.2020.106270.

[16] https://ml-cheatsheet.readthedocs.io/en/latest/logistic_regression.html.

[17] Feng Zhou, Tao Chen, and Lei Baiying, "Do not forget interaction: Predicting fatality of COVID-19 patients using logistic regression", arXiv:2006.16942v1 [stat.AP] 30 Jun 2020.

[18] www.javatpoint.com/linear-regression-in-machine-learning.

[19] Ramjeet Singh Yadav, "Data analysis of COVID-2019 epidemic using machine learning methods: a case study of India", https://doi.org/10.1007/s41870-020-00484-y.

[20] Raju Vaishya, Mohd Javaid, Ibrahim Haleem Khan, Abid Haleem, "Artificial intelligence (AI) applications for COVID-19 pandemic", https://doi.org/10.1016/j.dsx.2020.04.012.

2

Application of 3D Printing in COVID-19

M. Anantha Sunil, T. Sanjana, Akshata Rai, and Apoorva G. Kanthi

CONTENTS

DOI: 10.1201/9781003126218-2

2.1 Introduction

In 2019, a sudden disease started spreading throughout Wuhan, China. The most common symptoms were fever, cough, and tiredness. The symptoms were very similar to that of a common flu, yet no kind of medicine was making the affected patients feel better and before long, the disease took over the patient's body, causing major respiratory problems. It soon started spreading all around the world. This was the coronavirus, also called COVID-19. The doctors, scientists, and experts around the globe came together to understand what could be done to stop it. Many medicines were made, many ayurvedic and supplement tablets were advised, but none of them cured the disease. Soon the aim went from curing the disease to preventing it, because people started accepting that if one has great immunity one has the ability to fight this disease. Old people and people with underlying diseases, especially respiratory related were advised to stay indoors and to take extreme care of themselves. People having serious heart diseases, cancer, chronic disease, kidney problems, and those suffering from asthma, liver disease, and high blood pressure were advised to take extra precautions. By March 2020 many people were wearing masks and carrying sanitizers. Social distancing had become the new trend.

It was declared as a pandemic by the World Health Organization on 11 March 2020. [1] More symptoms were added to the list, such as shortness of breath, sore throat, chills, runny nose, and muscle aches. The world was not ready for such a widespread disease. The people were suffering, and there was not much that could be done to stop it. People from all professions came together to fight coronavirus and to make sure that it would not affect too many people, and that is when the world realized that it was lacking in so

FIGURE 2.1
A 3D model of the COVID-19 virus strain. [2]

many essential things required to deal with the pandemic. Figure 2.1 shows the 3D model of the COVID-19 virus strain [2]. Masks, face shields, gloves, hospital equipment, ventilators – there was a shortage of almost all these things. Eventually people found various ways to cope with the shortage. One of those ways was 3D printing. It helped in producing masks, hospital appliances, medical tools, and almost all the basic requirements to prevent and treat COVID-19. 3D printing is a very popular technology that has found application in every possible sector, be it architectural, medical, or even the food industry.

Products developed using 3D printing have several advantages. There is minimum to no wastage in the production of products, making it exceptionally efficient and quite eco-friendly. There are various types of 3D printing, which means that every purpose has its own technique, making the accuracy more precise.

2.2 Types of Modeling

There are various types of 3D printers. Each serving a specific purpose. Apart from being very specific, they are also divided into various types on the basis of their geometry, cost, and mechanical requirements/properties.

2.2.1 Fused Deposition Modeling (FDM)

Among various other types of modeling, fused deposition modeling and fused filament fabrication (FFF) are most widely used methods in 3D printing. The process of FDM is quite simple. Acrylonitrile butadiene styrene (ABS) is used as a printing material in this process. This modelling begins as a CAD (computer-aided design) file. The file needs to be saved as an .STL format so that the printer can decode and go ahead with the execution of instructions. The process starts with the nozzle moving up and down, basically horizontally and vertically in order to form a cross section. Every FMD modelling procedure will consist of two major materials, one used for the framework and the other, which acts as a scaffolding to support the object as it is being printed. So, once the printer has read and converted the file, it starts the execution.

The above-mentioned materials take the form of the model, which is basically a representation of various plastic filaments joined together. To start with, the plastic is melted and pushed out to the base with the help of the nozzle. The horizontal and vertical cross section get ready by following the coordinates that the base as well as the model maintain. This hot plastic cools down and sticks to the layer beneath it. This process is followed until the whole model is ready. As this is done layer by layer, a process called additive manufacturing [3]. Once a layer is ready, it is pushed down to make space for the next layer. The correct dimensions are met by the movement of the base and the nozzle, which is inputted in the computer itself. The time taken to make models depends on the size, dimensions, and accuracy. 3D printing a model takes more time if it is geometrically complex. When compared to other methods, this process is considered to take a longer time [4].

2.2.2   Selective Laser Melting (SLM)

The technique of selective laser melting started in 1995 at an institute in Germany. It is also known as direct metal laser sintering (DMLS). This name is not exactly suitable, as the real process happens due to melting and not sintering, which means the part is fully dense. This is very similar to the other selective melting processes.

The DMLS process uses various types of alloys. Since this process involves production of objects layer by layer, it allows for the creation of complex designs that otherwise would not have been possible. The final products are strong and durable objects that can be used as both functional prototypes and as end products. The process includes a high-power density laser to help in melting and fusing of the essential powders.

First, the 3D CAD file is sliced into various layers. These layers represent the 2D images of each layer. These images are later fed into the machine and the process starts. In the SLM process the thin layers of atomized metal powder are distributed on the substrate plate which is usually a metal. The 2D image is fused on the layer by selective melting of the metal powder. This process is repeated for all the succeeding 2D images until the very last layer in order to finalize the product with maximum accuracy. The procedure and the components involved are as shown in Figure 2.2 [5].

In this technology the metal powder is fused on the solid layer through selective melting of the powdered metal, locally using a high-powered beam. The parts of the objects are built layer by layer. Each layer is usually 20 micrometers thick. Some of the materials used in the process are Ni-based superalloys, tool steel, aluminium, stainless steel, cobalt chrome, copper, titanium, and tungsten. Tungsten (wolfram) has a high melting point. At low temperatures wolfram that would be ductile becomes brittle; this is basically the ductile-brittle transition. Thus, tungsten parts are produced by this method. For the metal to be included in this process it is necessary for it to be in powdered/ atomized form. This process is best suited to produce objects with complex structures, thin walls or which have voids, channels, and so forth. Industry applications include aerospace, manufacturing, medical, prototyping, and tooling.

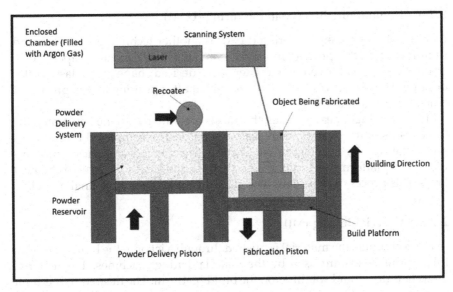

FIGURE 2.2
The various components and the way the process is carried out. [5]

2.2.3 Electron-Beam Melting

Another type of rapid prototyping is the technique of electron-beam melting (EBM). It is mainly used for metals. The material used for this process is (i) metal powder and (ii) wire.

i. Metal powder-based systems: In this the metal powder is consolidated into a solid by simply melting it. The raw materials are placed under a vacuum and then fused together, layer by layer, to produce parts. The heat in this process is provided by the electron beam, which is further computer controlled. This is different from selective laser sintering as the raw materials are melted for the action of fusing. This process takes place in the presence of little to no matter, making it suitable for forming parts of reactive materials like titanium, which has a high affinity for oxygen.

ii. Metal wire-based systems: With this, a part can be built by melting welding wire onto a surface using an electron beam. This is similar to fused deposition modelling. But here metal is used instead of plastic. The power source used in this process is the electron beam. The electron beam is quite efficient as it is more accurate and can be deflected as well. Electromagnetic coils (up to 1000+ hertz) support the deflection. Some of the alloys that can be used in this process are stainless steels, nickel alloys, titanium alloys, copper-nickel alloys, tantalum, cobalt alloys, and so forth. These can be found in the form of welding wires.

2.2.4   Laminated Object Manufacturing (LOM)

Laminated object manufacturing (LOM) is another technique used for rapid prototyping. In this process, multiple layers of adhesive-coated paper, metal, or plastic are glued together and later cut to desired shape with a laser cutter or knife. The objects printed can be modified after printing by the process of machining or drilling.

First, by using a heated roller, the sheets are adhered to the substrate. The dimensions of the prototype are traced by laser. The completed layer moves out of the way by moving down. Then the fresh layers are placed into position. The platform next moves into the position, ready to receive the next layer. This process is repeated to prepare the prototype or the final model.

2.2.5   Material Jetting (MJ)

In this technique, the material is pushed through a jet onto the build platform. Inkjet print heads are used by the material-jetting machines. The material then cools down and solidifies on the build platform. The desired part is built layer by layer. The finishing and accuracy can be achieved by this process, but only wax-like materials can be used for it. And, because of this, the part is rather fragile. Post-processing involves the removal of the support structures either mechanically or by melting it away.

Prototypes that are built by this technique are used for visual and form-fit testing. Casting patterns that are produced have very good accuracy and surface finishing. They are often used for lost wax casting in various industries, especially in the dental and medical fields. It is quite popular in the jewelry market as well [7].

2.2.6   Stereolithography (SLA)

This technique is also called resin printing or optical fabrication. The whole process revolves around photopolymer resin and ultraviolet (UV) laser. Stereolithography starts as a computer-aided design and, once that is prepared, it is converted into an .STL file. The file is now ready to be used to generate a model. The model is prepared on a build-platform, or base. This base is fully immersed in liquid resin. When UV light falls on the resin it starts to solidify in the area where the ultraviolet light falls. The UV light only falls on the surface with respect to the dimensions fed by the manufacturer. This process is done layer by layer.

As one layer is completed, the surface is smoothed out with the help of a levelling blade. The main task of the levelling blade is to coat the newly produced product with another layer of resin so that the UV laser can act on it. This process continues until the final product is obtained. This layer-by-layer process is quite selective and precise, as it follows the dimensions given. This process of coating, recoating, and solidification goes on until the end product is achieved. The model is then raised upwards from the liquid resin with the help of the build-platform. Cleaning and removal of extra material is done and, finally, this model is kept in an UV chamber to completely solidify. Apart from being quick this process also gives a really nice finish and is highly accurate. The only limitation is that it is expensive.

2.2.7 Binder Jet

Binder jet 3D printing can also be termed a power bed inkjet. The data in the CAD file is used to create the desired object. This technology was first developed at the Massachusetts Institute of Technology and patented in 1993.

As in many other additive manufacturing processes, the object, or the part to be formed, is printed layer by layer. The head of the inkjet pin moves over the powder bed, and, as it moves across it deposits the binding liquid over the bed. Later, a thin layer of powder is spread over this binding liquid and adheres to the surface. On this, again, the binding liquid is deposited, and the powder is spread, and the process continues with each layer adhering to the previous layer, until the desired object is formed. When the object is formed the unbound powder is removed automatically and/or manually. This process is called de-powdering. This unbound powder can be reused again to some extent. After de-powdering, the object can be subjected to other treatments to get the desired final object.

For the powder bed, starch and gypsum plaster and used, and the liquid binder may be water, which activates the plaster. The binder may include dye and other additives to adjust the viscosity, boiling point, and surface tension, as per the object's requirements.

The printing process is relatively quicker as compared to other additive manufacturing technologies. But the parts printed by this method are more porous and also have unfinished surfaces, as in other additive manufacturing methods that involve the melting of the metal. In the binder-jet printing method, the powder bed is not physically melted, but is joined by a binding agent. Since we use binding agents, it allows us to use materials, like ceramic, which have high melting temperatures, and materials like polymer which are heat sensitive. But the post-printing processes may require more time than that spent printing the object itself. The post-printing processes may include curing, sintering, and additional finishing.

2.3 Components of a 3D Printer

There are various types of 3D printers and each one is made up of various components. Some printers might have UV lasers, some might not. The components of any device depends on the functions it is required to perform. In this context we can read about some of the basic components. These components are present in almost all kinds of printers, although some might vary. The most important component is the controller board, also known as the brain of the printer. The controller board is responsible for the core operations, and every operation depends on the functioning of the controller board. The two next important components are the filament and frame. The printer can be used as a single or dual extrusion system according to its needs. The only main difference among the two is that the single dual system uses one kind of filament whereas dual, as the name suggests, uses two kinds of filaments. The material used to print the objects is called the filament, and the components and pieces are held together by the frame. This provides durability and stability. Further, the power supply unit (PSU) supplies the power to the entire printer. Apart from these a set of components is required to control the motion and to create motion in the printer as well. The motion in direction specified – X, Y, Z dimensions to be precise. The movement of all the axes is controlled by a motor known as the stepper motor. Every axis has a special component devoted to it to further create motion in that direction. The movement in the Z-axis depends on the threaded rods, whereas the movement in the X and Y axis depends on the belts. Another component is part of the set of motion components to prevent the axes from moving beyond their dimensional range. They are treated as markers and are called end stops.

The components involved in the processing of the object are print bed, print bed surface, and the print head. Now, to create an object with strict dimensional range, these components are quite important. The print bed is where the filament is deposited to form the required product. Further, the print bed surface goes on top of the print bed in order to ease the process of removal of the object from the surface. Finally, the print head does the unique work of creating the object. The object is created using the filament, which is melted and then passed through the nozzle. This nozzle is attached to the print head.

2.4 Materials Used

Some of the materials used are polylactic acid (PLA), photopolymers, and polyvinyl alcohol (PVA), high-impact polystyrene (HIPS), high-density polyethylene (HDPE), polycarbonates (PC), polyamides, and acrylonitrile butadiene styrene (ABS). Steps involved in 3D printing include part design, converting to STL file, transferring file to the machine, setting up the machine, building, post-processing and then using it for a certain application [8].

2.5   Applications of 3D Printing in COVID-19

3D printing has been used extensively in the department of medicine. Its popularity has been seen to only increase with time. The never-ending applications of 3D printing in the medical world is astonishing and mind-boggling. It has been used to make hearing aids, models of organs, prosthetics, replacement limbs, implants, and vascular organs. Apart from this it has long been used to make surgical equipment. The era of 3D printing will go down in history for the very relentless and smooth preparations of medical devices. It has proven to have also contributed to making existing equipment even better.

Medical tools are available faster, in more bulk, and at cheaper rates after the initial investment in a 3D printer. Things like custom-made prosthetics, which in the past could only be dreamed of, are now being made and used worldwide. This technique of additive manufacturing can be used on casts that are won by people with fractures, which is quite common, making the demand for casts directly proportional. Despite the advantages of 3D printing products – being custom made, easy to make, and time efficient – they are also cheaper when compared to comparable equipment. These applications are not limited only to this but, somehow, they found their way to help tackle COVID-19.

The main precautions that could be taken to prevent the spread of COVID-19 is to wear a mask when outdoors, use of gloves, keeping track of immunity, and maintain social distancing. The need for things like masks, gloves, face shields is increasing by the day. If a person does not have a strong immune system or comes in contact with someone suffering from COVID-19, that person can become infected as well. The treatment after contracting COVID-19 requires ventilators for patients who have acute illness. The science and medical industries yet again have come together, with the help of additive manufacturing, to deal with this problem. The application of 3D printing in COVID-19 is segregated into six major categories: Emergency wards, visualization and practice aids, testing devices, personal protective equipment (PPE), medical devices, and medically complimenting devices such as mask adjusters. You will read about these domains in the following sections.

2.5.1   Nasopharyngeal (NP) Swab

As the infection of COVID-19 spreads, the need for conducting tests increases. The demand for COVID-19 test kits is very high, and the nasopharyngeal swab is an integral part of the kit. Thus, the demand for nasopharyngeal swab production has increased rapidly. The shortage results in the reduction

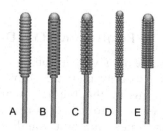

FIGURE 2.3
A few of the 3DP swab designs. [9]

of testing, thereby creating a situation so compromising that it could prove fatal to many.

First, the researchers came up with a single-part design. Later, in the clinical trials, the 3D printed swab was tested against the traditional swab. The traditional swab already had the virus strain from a coronavirus-positive patient. The results were satisfactory as it showed that the 3D printed swab performed well compared to the traditional swabs and, in a few cases, they performed better. In the beginning, around twenty swab designs were made in order to test and find the very best one. The design satisfying the minimal conditions of surface area and comfort was selected as the best. Few designs can be seen in Figure 2.3 [9]. The tip design that is currently in use is "Tip C". To make sure that the tip designs picked up enough of the sample for the viral testing, bench lab testing was conducted.

Swabs are printed with their bases directly in contact with the platform without any rafts or supports. To rinse 3DP swabs, a form wash in 99 percent isopropyl alcohol for 20 min is undertaken. They are washed while still attached to the build plate. Later they are air dried for 30 min and then scraped off gently from the build platform. Then the swabs are suspended in a curing rack with their tips pointing down, and are then placed in the Form 2 cure at 60 °C for 30 min or in Form 3B at 70 °C for 30 mins. After the completion of the curing, 3DP swabs are prepared for sterilization by placing them in steam sterilization pouches.

After they are sterilized, they should be prevented from becoming more brittle over time, by keeping them away from direct light. To complete the COVID-19 testing kit, after sterilization the swabs are individually packaged and then packed with the test tube and viral transport medium (VTM). These kits are later distributed to the testing centers for use [9].

2.5.2   Face Shield

COVID-19 is an infectious disease, and health workers are selflessly working day and night to take care of patients, to keep the people safe. It is very important for health workers to protect themselves from the virus. They require personal protective equipment like face masks, face shields, and gloves to protect themselves from getting infected. The face shields help health workers protect themselves from bodily fluids and droplets from the patients who are being treated for cough, cold, and other infections that are symptoms of COVID-19. There is a high demand for the PPEs, and it is difficult to meet the requirements through the conventional methods of production. The fabrication of face shields by 3D printing technology increases the production rate.

The 3D printed face shields include three components – a headband, a shield and an elastic strap. The two arcs in the headband are for the purpose of supporting the forehead and the to deflect the shield away from the face so as to avoid fogging. For this product, fused deposition modelling (FDM) is used. The materials include polyethylene terephthalate glycol (PETG) or acrylonitrile butadiene styrene (ABS) and polylactic acid (PLA). The printing takes around three hours. Since PLA is biodegradable, it has low environmental impact. It costs $1–1.10 per piece of the headband. Transparent PVC sheets, which are A4 size sheets and 0.2 or 0.3mm in thickness are used in shields. These are perforated and then horizontally attached on the anti-return spikes. For the elastic straps, they are cut from a tourniquet or a buttonhole elastic roll. Once these parts are put together, they are checked for quality. They are checked for the flexibility of the two arches and for the mechanical resistance of the positioning pins for the shield and the elastic band. And the edge and surfaces are smoothened thus rectifying the shape defects. They are disinfected by keeping them in sodium hypochlorite at 0.5 percent for 15 minutes. Before they are sent to the hospitals the shields are tested in realistic conditions. In the first week, 1,547 face shields were produced, 2,989 in the following week, steadily increasing to reach a total of 10,151 at the end of 5 weeks. After use, face shields have to be disinfected. This can be done by placing the headband in sodium hypochlorite (0.5%) for 15 minutes. The shield can be changed, and the elastic strap can either be changed or washed at 60°C.

The production of face shields is done in various stages. First, an interdisciplinary group was created from academics involving technicians, engineers who are practitioners, and the unfulfilled medical needs are identified and clarified. Various types are prototyped and manufactured for the practitioners. Mass-produced solutions are designed and pre-industrialized, while iterating on other possible solutions. Then the mass production of approved solutions is undertaken [10].

2.5.3 Various Masks

There are several kinds of masks used for Covid-19 and which can be manufactured using 3D printing technology.

2.5.3.1 N95 Masks

During these times, healthcare workers need to ensure that they are following all sorts of precautions. The main aspect being the use of professional masks. The fit and comfort of the masks can be improved by using 3D printing to print tailored seal designs. In Figure 2.4 [12] the mask is shown in different views. The facial parameters can be scanned using 3D laser scanning, so as to customize face mask seals. Customization is done on the basis of the buyers' facial structures, such as their face width, jawline, nose structure, and so forth. In a study using face seal prototypes with acrylonitrile butadiene styrene plastic using a Fused Deposition Modeling 3D printer, three subjects showed improved contact pressure. Electrostatic nonwoven polypropylene (PP) fibers are lightweight and inflexible. This material is used as the filtration material in the standard N95 masks. After cooling, the 3D printed parts may be distorted due to the semi-crystalline structure, making the process difficult. A blend of polypropylene (PP) and styrene-(ethylene-butylene)-styrene (SEBS) provides better flexibility and also better printability for the N95 mask design. PP is widely used in various industrial applications not only for its low cost, but also for its printability, mechanical integrity, processability, and recyclability. Since, SEBS is a synthetic polymer with elastic properties and low distortion during extrusion and low processing temperatures, the combination of PP and SEBS improves the fundamental texture of the masks. By controlling the thermoplastic elastomer ratio, the flexibility and elasticity of the 3D printed N95 masks can be altered. To print the masks, which are biocompatible and stable in comparison to the industrial manufacturing brands, different methods can be used. These include 3D melt electrospinning

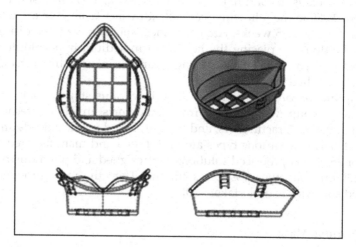

FIGURE 2.4
The mask in different views. [12]

printing, which can be used to create PP microfibers with sequential layering to accurately obtain a 3D form [11].

Contoured surfaces are the main component that makes contact with the user's face. On the exterior side there is a 60mm x 60mm hole where a 60mm x 60mm filter holder grid is placed. Most of the fabric is used as the filter material. The mask was designed using CAD software. A variety of filaments were used, such as thermoplastic polyurethane (TPU), polylactic acid (PLA), and polyethylene terephthalate glycol (PET-G) [6].

For better seal and more comfort there is an addition to the masks: liquid rubber. This step is optional. It requires submerging the PLA in hot water (around 60°C) for a few seconds making it appropriate to form a mould adjustable to the face. As much as wearing a mask is important, the fitting of the mask is equally important. The fitting of the mask increases its efficiency. Here we discuss ways in which this problem can be addressed.

2.5.3.2   Snorkel Mask Adapter

An already available snorkel mask is made to combine with an adapter to form the snorkel mask adapter. This functions as a full-face respirator. The process starts with a CAD file that is further converted to an .STL file, making it suitable to obey the given instructions. Various filaments are used for printing, including polyethylene terephthalate glycol (PET-G), thermoplastic polyurethane (TPU), and polylactic acid (PLA). We chose PLA mainly because of its melting point. It is also easy to work with and proves quite cost effective [13].

To identify leaks in the design, the subject wore the mask, adapter, and viral filter and was submerged in water. The water level was maintained above the point of connection of the filter to the adapter, but it was maintained below the top of the filter. The results of the fit testing were satisfactory for the use of this device as an alternative to the N95 masks.

2.5.3.3   3D Printed Mask Frames

The use of these mask frames can improve the lifespan of the N95 and KN95 masks. They can also act as a substitute for defective or damaged bands without affecting the fitting. They also help in improving the fitting. The mask was designed in Rhinoceros (R) Rhino 6 and the print settings were the default values for the 3dPrinterOS customized for the 311 Dremel 3d45 3D printer. The printing time for each regular-sized mask frame is around 30 minutes. For the assembly, two methods are used, with each involving a slight modification to the printed lateral frame. In method 1, to join the frame components a fast-acting adhesive like cyanoacrylate is used. In method 2, the connection was purely mechanical, where wires were twisted to join the components of the frame. The attachment of the mask to the respirator employs clips. After assembly the frames have to be sterilized by using 70 percent isopropyl alcohol wipes [14].

2.5.3.4 Mask Extenders

This pandemic has forced everyone to wear a face mask in order to stay safe and as a precautionary measure. At the height of the pandemic, everyone from the general public to the working class was expected to wear it. The health workers need to wear the masks for longer periods of time. A prolonged period of wearing the mask causes discomfort to the wearer due to heat, pressure, and pain. The pain is caused by the mask design, which involves an elastic strap looping around the posterior auricular region, causing a constant pressure and friction against the skin. The purpose of mask extenders is to ease the discomfort. Initially an already available design was used. This was paired with the surgical masks, which have elastic straps. Later the design had to be changed, since the initial design was not suitable for everyone.

The slicing software used was Ultimaker Cura. 3D printing was performed with Ultimaker SS and CR-10. The print settings utilized 20 percent of Infill, a basic print speed of around 55–50 mm/s, 64° Celsius for the built plate and a moderate 215° Celsius at the nozzle, which targeted the first layer, the preceding layers maintained at 21° Celsius [15]. There was no requirement for support or raft. This new modified design does not allow for the production of high numbers of masks in a short period of time. Disinfectant sprays are used to further disinfect the extenders. The final design was printed on two 3D printers with 7 straps being printed on each of the printers. Around 457 successful mask extenders were produced and delivered.

2.5.3.5 Open-source 3D Printed Ventilator Device

Ventilators play a very crucial role during this time of pandemic. The need for ventilators around the world steadily increased since the beginning of the pandemic. Here the discussion is about the open-source 3D printed ventilation device, which is capable of controlling the breathing rate, volume, and pressure and is designed according to the observations made in the COVID-19 hospital wards. This device is based on the principles of hydrostatic pressure, using three containers of water. The designed setup does not involve the use of pistons or gearboxes as in traditional devices. Thus, the smaller number of moving elements eases the sterilizing and the reuse of the equipment between patients. The setup requires an electric motor, and this motor is retained in the motor housing, which was 3D printed. The rotational movement of the electric motor has to be converted to linear motion so as to linearly move the container. The height regulator is 3D printed. There is a requirement of a 3D printed pulley system in this design setup to reduce the overall number of parts needed. All the parts were 3D printed on an Anycubic Mega using polylactic acid filament. The containers that are 3D printed are designed to be watertight to avoid water or air leakage and also eliminating the need for the adhesives, thus reducing the chance of leakage [16].

2.5.3.6 Hospital Respiratory Apparatus

This pandemic was a wakeup call for many production sectors as well. During this period the shortage of important equipment was realized, and the problem was faced head on by the various techniques of 3DP. Key equipment, such as masks, hoods for ventilation systems, and venture valves, had to be produced in abundance to meet the demands.. The United States Food and Drug Administration (FDA) does not object to the creation and use of devices such as T-connector that meet specifications described in the instructions provided to the FDA for use in placing more than one patient on a mechanical ventilator when the number of patients who need invasive mechanical ventilation exceeds the supply of available ventilators and the usual medical standards of care have been changed to crisis care in the interest of preserving life. The FDA's "no objection" policy applies during the duration of the declared COVID-19 emergency. The creation of 3D ventilator splitters and adjustable-flow control valves, like the no2covid-ONE valve, allows the use of a single ventilator for multiple patients who have different oxygen requirements [17].

2.5.3.7 3D Printed Isolation Wards

At its peak, the infection rate of COVID-19 was very high, and thus the number of infected people increased day by day all over the world, and the hospitals and quarantine centers were running out of space to accommodate the patients. In Xianning, China, additive manufacturing paved a way to solve this problem. Winsun has 3D printed isolation houses, which eases the burden on hospitals and quarantine centers. Sectors such as construction have been using 3D printing for a long time. This technique has lots of advantages: it is faster, and the requirement for raw materials is less than in conventional methods. This also allows for creating temporary dwellings when in crises, and these can be placed in any area where they are required. The mobility of these structures is their biggest benefit. These are deployable in any location. According to Winsun, in 24 hours they were able to 3D print 15 houses, which means they spent less than 2 hours for each house. These houses are not only meant for patients but also for medical staff. According to Winsun, each house can accommodate two people. They have built-in showers, air conditioning, and toilets, and they meet quarantine standards. The durability of the house is increased by the use of environmentally friendly materials [18]. The rooms are printed with the material that is a combination of concrete and recycled materials [19].

2.5.3.8 *Contact-free Devices*

COVID-19 infection can be transmitted through direct contact and via objects or surfaces such as door handles, lift buttons, and so forth in public places, including medical centers. The risk of transmission can be reduced by meticulous and regular surface cleaning of the objects. But this only partially addresses the issue. Another way is to avoid skin-to-surface contact. Here, the discussion is about various 3D printed objects that may help to avoid the skin-to-surface contact.

2.5.3.8.1 *Door Openers*

A design was adapted [20] for a door opener with clips and cable ties on straight and circular sectioned handles. This design also uses cable ties on an elliptical sectioned curved handle. FDM devices were used to print these door openers. The printing characteristics of the door openers is as follows: the layer thickness was 0.33mm, the filling density 43 percent, and the wall thickness 2mm.

2.5.3.8.2 *Door Hooks*

Some door handles have very different designs, and it is not possible to design one door openers for all types of handles. For such designs a door hook has been designed that is protected by a retractable sheath. The printing characteristics of the door hook and the sheath is as follows: their filling density is 100 percent, printing time was 90 minutes, the hook's length was 80mm and its thickness 5mm.

2.5.3.8.3 *Button Pushers*

The components of a button pusher include a cylindrical tube and a retractable tip. The tip can further be blocked by a pressure pin. This geometry cannot be printed using FDM since there is a gap that has to be printed between the tube and the tip. Thus, Polyjet J735 and J750 printers were used in producing this device [21]. The end product was an already-assembled button pusher. The printed button pushers had to be treated post printing using waterjets to remove support material. Later were are cleaned using a solvent before dispatching them.

2.5.3.8.4 *Hand Sanitizer Grippers*

It is very important to disinfect ourselves frequently by washing our hands or with the use of hand sanitizers. There are sanitizers kept in public places, too, but what we tend to ignore is the fact that the sanitizer bottle itself can be a medium for the spread of the virus. A possible solution to this problem can be the use of 3D printed wrist clasp which can safely hold the sanitizer bottle. The wrist attachment allows the user to apply the sanitizer gel onto the palm without actually having to hold the bottle, thus, protecting the user from the spread of the virus from the potentially contaminated bottle. Touch-free hand sanitizer dispensers can also be 3D printed. The user need not touch the dispenser. The hand will be sanitized if the user gets his/her hands closer to the dispenser.

2.5.3.9 Drone Parts

During a time of pandemic, it is very risky to go out in public places even to buy medicines, food, groceries, and so forth, and due to the lockdown, the supply of essential needs is reduced. As a solution to this problem, drones can be used to reach the worst-pandemic hit areas instead of sending people there. These drones are used for safety and communication purposes. They can scan the suspected areas and can send an automatic message for the necessary actions to be taken. These 3D printed drones have been equipped with mounts, extension arms, hooks or payload containers, and so forth, with necessary components, and they can stay in air for almost an hour and carry loads up to 13kg. These drones are used for delivery of foods and medical supplies. They can also be used to disinfect pandemic hit areas without the need for human intervention in a very short amount of time.

2.5.3.10 3D Bioprinting

Another way of predicting toxicology and to model pathologies was seen in a company in Russia. They are also being used for research purposes to come up with drugs for the COVID-19 disease [22].

2.5.3.11 Antimicrobial Polymers in the COVID-19 Pandemic

Antimicrobial polymer controls the growth of microorganisms such as protozoans, fungi, and bacteria. Antimicrobial polymers help in the development of critical medical devices [22].

2.6 Conclusion

This chapter revolves around the various ways in which 3D printing has lent a hand during this pandemic. Without 3D printing we would not have been able to achieve multiple products. Making products is one thing, but making a product which is cost effective, time efficient, and user friendly is what was made possible by the printer. The power to modify and convert existing items into the best version of itself should be the main aim of any manufacturer, and it has been made possible by this technique of additive manufacturing. This chapter explained various techniques used to make different products, such as door handles, masks extenders, door hooks, sanitizer grippers, and drone parts.

A very important aspect is yet to be covered by 3D printing, and this deals with medicines as a whole. Developing oral dosage syringes using 3D printing is of greater interest. The syringe should be made using the 3D printer, and it should allow patients to treat themselves at home. When patients who home quarantine themselves can give themselves the right medication will help overcome many difficulties. A drug can be administered, and the syringe will make sure that the dosage is accurate. The world of 3D printers has helped fight COVID-19 in the best possible way.

References

[1] www.who.int/emergencies/diseases/novel-coronavirus-2019/events-as-they-happen (Accessed on 6 October 2020).

[2] Tino, R., Moore, R., Antoline, S. et al. COVID-19 and the role of 3D printing in medicine. *3D Print Med*, 6, 11 (2020) 2.

[3] www.livescience.com/39810-fused-deposition-modeling.html#:~:text=There%20are%20several%20different%20methods,create%20a%20three%20dimensional%20object (Accessed on 6 October 2020).

[4] Sai, P. Chennakesava and Yeole Shivraj, Fused deposition modeling – insights. International Conference on Advances in Design and Manufacturing (ICAD&M'14) (2014), pp. 1345–1350.

[5] Jiao, Lishi, Chua, Zhong, Moon, Seung, Song, Jie. Bi, Guijun, and Zheng, H, Femtosecond, Laser produced hydrophobic hierarchical structures on additive manufacturing parts. *Nanomaterials*, 8(8), 601 (2018).

[6] Azam, Farooq, Abdul-Rani, Ahmad-Majdi, Altaf, Khurram, Rao, T. V. V. L. N. and Zaharin, Haizum, An in-depth review on direct additive manufacturing of metals, IOP conference series: Materials Science and Engineering (2020).

[7] www.additively.com/en/learn-about/material-jetting#:~:text=Process%20description&text=Material%20jetting%20machines%20utilize%20inkjet,are%20used%20with%20this%20technology. (Accessed on 6 October 2020).

[8] Zachary O'Connor, Daniel Huellewig, Peeti Sithiyopasakul, Jason A. Morris, Connie Gan, and David H. Ballard, 3D printed mask extenders as a supplement to isolation masks to relieve posterior auricular discomfort: An innovative 3D printing response to the COVID-19 pandemic. *3D Print Med*, 6, 27 (2020).

[9] Ford, J., Goldstein, T., Trahan, S. et al. A 3D-printed nasopharyngeal swab for COVID-19 diagnostic testing. *3D Print Med*, 6, 21 (2020).

[10] Lemarteleur, V., Fouquet, V., Le Goff, S., Tapie, L., Morenton, P., Benoit, A., Vennat, E., Zamansky, B., Guilbert, T., Depil-Duval, A, Gaultier, A.L, Tavitian, B., Plaisance, P., Tharaux, P.L., Ceccaldi, P.F., Attal, J.P., and Dursun, E. 3D-Printed protected face shields for health care workers in Covid-19 pandemic. *Am J Infect Control.* 9(3) (2021 Mar), 389–391 , 4. doi: 10.1016/j.ajic.2020.08.005. Epub 2020 Aug 11. PMID: 32791260; PMCID: PMC7417271.

[11] Ishack, S, Lipner, S.R. Applications of 3D printing technology to address COVID-19-related supply shortages. *Am J Med.* 133(7) (2020), 771–773. doi:10.1016/j.amjmed.2020.04.002.

[12] Dalla, S., Bacon, B., Ayres, J.M., Holmstead, S., Ahlberg Elliot, A.J. 3D-printed N95 equivalent for personal protective equipment shortages: the Kansas City Mask. *J 3D Print Med.* 2021;10.2217/3dp-2020-0019. doi:10.2217/3dp-2020-0019.

[13] Dalla, S., Shinde, R., Ayres, J., Waller, S., and Nachtigal, J. 3D-printed snorkel mask adapter for failed N95 fit tests and personal protective equipment shortages. *J 3D Print Med.* 2021;10.2217/3dp-2020-0018. doi:10.2217/3dp-2020-0018.

[14] McAvoy, M., Bui, A.N., Hansen, C., Plana, D., Said, J.T., Yu, Z., Yang, H., Freake, J., Van, C., Krikorian, D., Cramer, A., Smith, L., Jiang, L., Lee, K.J., Li, S.J., Beller, B., Short, M., Yu, S.H., Mostaghimi, A., Sorger, P.K., and LeBoeuf, N.R. 3D printed frames to enable reuse and improve the fit of N95 and KN95 respirators. medRxiv [Preprint]. 2020 Jul 26:2020.07.20.20151019. doi: 10.1101/2020.07.20.20151019. PMID: 32743606; PMCID: PMC7386530.

[15] O'Connor, Z., Huellewig, D., Sithiyopasakul, P. et al., 3D printed mask extenders as a supplement to isolation masks to relieve posterior auricular discomfort: an innovative 3D printing response to the COVID-19 pandemic. 3D Print Med 6, 27 (2020) 2–4.

[16] Faryami, Ahmad and Harris, Carolyn. (2020). Open-source 3D printed Ventilation Device. 10.1101/2020.05.21.108043.

[17] Tino, R., Moore, R., Antoline, S. et al. COVID-19 and the role of 3D printing in medicine. *3D Print Med*, 6, 11 (2020). https://doi.org/10.1186/s41205-020-00064-7

[18] www.3dnatives.com/en/winsun-coronavirus-260220205/ (Accessed on 6 October 2020).

[19] www.3dprintingmedia.network/winsun-3d-printed-isolation-wards-coronavirusmedical-workers/ (Accessed on 6 October 2020).

[20] François, P.M., Bonnet, X., Kosior, J., Adam, J., and Khonsari, R.H. 3D-printed contact-free devices designed and dispatched against the COVID-19 pandemic: The 3D COVID initiative. *J Stomatol Oral Maxillofac Surg.* 122(4) (2021 Sep), 381–385. doi: 10.1016/j.jormas.2020.06.010. Epub 2020 Jun 26. PMID: 32599093; PMCID: PMC7318987.

[21] Tay, Joshua. Cross, Gail, Lee, Chun, Yan, Benedict, Loh, Jerold, Lim, Zhen, Ngiam, Nicholas, Chee, Jeremy, Gan, Soo, Saraf, Anmol, Chow, Wai, Goh, Han, Siow, Chor, Lian, Derrick, Loh, Woei, Loh, Kwok, Chow, Vincent, Wang, De, Fuh, Jerry, and Allen, David, Design and clinical validation of a 3D-printed nasopharyngeal swab for COVID-19 testing, *medRxiv* (2020).

[22] Arora, Pawan, Arora, Ranjan, Haleem, Abid, and Kumar, Harish, Application of Additive manufacturing in challenges posed by COVID-19, Materials today: proceedings, 6–7 (2020).

3

Role of IoT and AI in COVID-19

Maligi Anantha Sunil, T. Sanjana, Akhil S. Raj, and A. Abhishek

CONTENTS

3.1 Introduction

COVID-19 outbreak has had a huge impact on human health and economic status of the entire world. It is essential to know certain key ways of understanding the disease so that the means to curb it can be found out. Researchers and scientists have come up with ways to avoid the spread of COVID-19. which include social distancing as well as sanitation after some or the other form of contact. Internet of Things (IoT) and Artificial Intelligence (AI) techniques are used to develop systems to fight this virus, and some of these have been implemented, and decisions have been taken, to reduce the impact of the pandemic. In this chapter various research works are compiled, and industry applications of IoT and AI to fight COVID-19 are presented.

DOI: 10.1201/9781003126218-3

AI is also called Machine Intelligence and can be broadly defined as a software which can sense, reason, act, and adapt. Technically speaking, AI is subdivided into Machine Learning and Deep Learning. These technologies and algorithms are extensively used – if not for this technology, the Covid-19 pandemic would have become even worse and would have impacted more.

Machine Learning (ML) is based on statistical mathematical models and probability, and this subdomain is a collection of algorithms classified under supervised (with labels) and unsupervised (without labels) and reinforcement learning (basically rewarding the agent for an activity done correctly). These algorithms help in forecasting and predicting COVID-19 related information, which helps in taking better decisions to combat this virus.

Deep Learning (DL) is actually sub-domain of both AI and ML, it includes "Neural Networks", which come in varieties such as Artificial Neural Networks, Convolutional Neural Networks (used in Images Primarily), and Recurrent Neural Networks (RNN – for sequence models such as speech recognition, natural language processing, etc.). It is easy to develop DL algorithms to predict whether the mass public is wearing masks or not, or whether they are maintaining social distancing; to develop algorithms to conduct radiographic analysis of the lungs, and much more.

The Internet of Things (IoT) describes the network of "things" embedded with sensors, software, and other technologies for the purpose of connecting and exchanging data with other devices and systems over the Internet. The latest technology of 4G and 5G WiFi has helped transfer data over the Internet and, with the advent of cloud storage and computing, the storage factors for these IoT devices help in reducing the cost as well as size and weight. IoT is now leading in the making of smartphones and automation. This key technology is used in drone systems where it can be used for no-contact delivery or for an automated soap dispensing system for no-touch handwashing, or the thermal temperature sensor used to let people into public places. IoT is widely regarded as a crucial tool to help in combating COVID-19 pandemic in many areas and societies. In particular, the heterogeneous data captured by IoT solutions can inform policy making and quick responses to community events.

3.2   AI and IoT for Large- and Small-scale COVID-19 Screening and Monitoring

As the old proverb says, "Prevention is better than cure"; one needs to understand that it is not possible to succeed at prevention of disease if the majority of the population is infected when there is no proper infrastructure, especially in underdeveloped countries. Hence, there are technologies such as IR Thermal detector, which helps in screening people before they enter public places. Research has been conducted on screening people on a large scale, and many works cite that thermal cameras can be used to detect the surface temperature of the person who is out in public, and issue a warning. Let us consider drones for a moment, since it is thoroughly established that COVID-19 is spread through some form of contact and, therefore, N number of drones can be used in the fleet and the capacity C (the number of tests that can be loaded onto one drone) are the variables (representing the resources) – the number of days D, needed to collect tests from a city's whole population, as the function of N and C. This can be a bit expensive but can be worth the money as it solves the nation's infrastructure problem. This helps in solving large scale no-contact delivery. Thermal cameras are used not only for measuring surface body temperature but also help in taking into consideration breathing rates and exhaled air temperature. Cameras are also used to monitor social distancing among the public to have better control over this crisis. These ideas are implemented in the following way:

Monitoring Social Distancing:-Deep learning is capable of detecting people and establishing a bounding box and giving IDs (provided privacy is not infringed upon). By the use of thermal cameras, people detected with high body temperatures can be localized, and the YOLOv3 algorithm helps in fast detection, identifying the person and calculating the nearby people and taking a note of it, and issuing a warning if close contact is found and is traced to need further treatment. The algorithms mentioned above are state-of-the-art algorithms with industry-ready implementations [1,2].

Thermal cameras: Data is collected from thermal cameras and analyzed using AI. Intelligent screening devices get the thermal and corresponding facial recognition videos. After getting the thermal videos, the first step is to extract respiration data from faces in thermal videos. During the extraction process, a face detection method is used to capture people's masked areas and then bidirectional RNN with GRU, and an attention mechanism is used to know the respiratory conditions in a sequential timely manner, and this sequence model helps in classifying the person as positive or negative [3].

Large-scale mass detection is possible with the right infrastructure and quick action with the help of thermal cameras, drones and Deep Learning algorithms. Technologies to monitor such as public queues are a possibility, or high precision thermal cameras for surface temperature detection through traffic lights can be used. When a traffic light is red for a particular intersection. then it does the scan around the people. FLIR ITS-Series sensors help in detecting heat of all objects in the scene, where it is fine-tuned to get the person's temperature. This technology can be used in airports, railway stations, and religious institutions [4].

There are autonomous Intelligent Care Robots developed by Vayyar and MediTemi, which has 4D imaging sensors that detect breathing patterns, heart rates, and fevers with the help of motion sensors and algorithms needed to monitor a room. It detects if a person has entered and then goes towards him and screens him, this is very efficient, as it keeps track of people who come in and stores all their readings and heart and breathing pattern waveforms [5].

There have been a lot of mobile apps that help in giving feedback of a person's health as a caution for him to be responsible for his surroundings. Example apps include: Stop Corona, a web-based application to issue hotspot alerts; nCapp, which has real-time cloud storage and accesses all the forms submitted by the people and creates a database to try to give accurate results based on the form the person has filled in. It also has a follow-up feature enabled [6]. Small-scale screening and monitoring are useful as no-contact systems for doctors to screen patients in clinics.

3.2.1 Quarantine Tracking

An app is developed for the purpose of tracking patients who are under home quarantine. This app helps in aiding a person who may have been in contact with a positive coronavirus patient by sketching the travel history. The app monitors the location of a person who may be corona positive and the user of the app.

If the travel history of both people match, the app helps to explain the symptoms of corona to the user and, if the symptoms match, it notifies the user to go to the nearest COVID-19 test center. This is how AI is used in this application.

The name of such of an app is Arogya Setu, which is developed by the government of India. It has the same features as described above, showing not only the patient's travel history but also the time and date. It helps in knowing the proximity of a person to the infected victims to stay alert and walk out of our homes with caution.

3.2.2 IoT Q-band

A low-cost, IoT-based wearable band to detect and track COVID-19 quarantine subjects is proposed in [7]. The World Health Organisation (WHO) has declared COVID-19 as a pandemic, and quarantine is playing an important role in containing its spread. A wearable makes a subject more compliant towards healthcare routines/restrictions; thus, an IoT-based wearable quarantine band is used (IoT-Q-Band) to detect any absconding. As mentioned above, the tracking of COVID-19 is through mobile applications and works on the concept of crowdsourcing. As per this app's user reviews, the functionality of alerts to nearby users is limited, since it assumes that every person who has a smartphone has this application installed. Though tracking through a mobile app has become inadequate, a visual indicator-based tracking method (e.g. medical authorities stamping on the hands with a non-washable ink) is most successful in detecting the absconding COVID-19 quarantine subjects. But it was found that in most of the scenario, the absconding subjects were reported from the public to be far away from quarantine location and moreover, this information was not notified to concerned authorities. Thus, to detect absconding subjects on time, none of the two mentioned schemes seems to work effectively, and they also somehow violate the privacy of subjects. Thus, a wearable band bundled with a mobile application is most useful as it performs real-time tracking. Figure 3.1 represents the system architecture of the IoT-Q-Band system.

The wearable band is worn by the quarantine subject on either hand or leg and is wirelessly connected to mobile application through a Bluetooth connection. The processing unit of the band repeatedly senses at specific time intervals, whether the band has been tampered with. After sensing, the band transmits the status (a byte of data) to the mobile application at two-minute intervals. Concerned medical authorities are responsible for registering the concerned person to IoT-Q-Band and also setting up the duration of the quarantine and authentication details. The supervisory role of the authority automatically rules out the concern for malicious data entering the tracking system and also preserves the privacy of quarantine subjects. During registration, along with personal details, the GPS coordinates of the quarantine

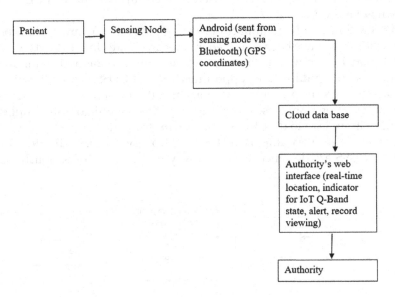

FIGURE 3.1
System architecture of the IoT-Q-Band system. [7]

location is stored. A designated person can monitor each registered quarantine case via a web interface and alerts are specified.

The IoT-Q-Band presents a minimalist design, keeping in mind the cost, global supply chain disruption, and the WHO's recommended average quarantine period. The IoT-Q-Band, with the bundled mobile application and cloud-based monitoring system, provides a scalable, low-cost solution to detect absconding COVID-19 quarantine subjects and track them in real-time. Due to the possibility of contamination, the reuse of an IoT-Q-Band could be avoided. Thus, the cost of this prototype is kept down by reusing several smartphone features. As an economical solution, the IoT-Q-Band could benefit low-income areas of the world where it could be used to keep track of quarantine subjects.

3.3   Clinical Decision Support System (CDSS)

In the present and also the future we will have to depend on IoT and AI technologies to effectively improve the healthcare sector. CDSS is a system for the managing as well as monitoring of patients in the ICU that uses IoT devices capable of both collecting and streaming physiologic data from ventilators, which helps in monitoring respiratory conditions and other medical devices. Machine-learning models are being embedded in CDSS that help to determine the condition of ventilators used in ICUs [8]. Workflow in CDSS is as shown in Figure 3.2.

CDSS will have a huge impact in treating patients who are in dire condition, where they have been stationed in Intensive Care Unit. Patients who are in ICU may have multiple ailments and need to be monitored continuously by doctors and multiple life support machines. If CDSS is used, it will help in correcting the respiratory conditions and the person can be under constant observation from these machines. But if these ventilators are configured wrongly, it can lead to disastrous situation for the patient.

Ventilators are very important for the recovery of a COVID-19 affected patient. CDSS are designed in such a way that data is continuously being

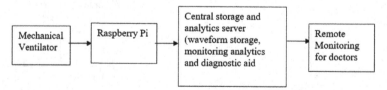

FIGURE 3.2
Workflow in CDSS.

collected from multiple concurrently operating mechanical ventilators. Whenever there are changes in the person's breathing pattern or any abrupt conditions, the sensors in the system will directly notify the doctor to take action. This works same as one of the devices that will come up later in this chapter – one that has the same principle of constant monitoring. In this system the sensors are connected such that pulsating reading of breath is obtained in waveform format, and these are kept in the cloud server for better storage and to have no strain on the device. Whenever physicians want to, they can access these data and make accurate decisions to help the patient in recovering.

A Raspberry Pi microcomputer, with a small customized linux-based system software, is connected to the ventilator through a network. After collecting data, ventilator waveform data (VWD) can be used by physicians to access the patient's files by means of a mobile application.

It is known how AI systems can be trained to recognize data, analyse it, and process it along with having the ability to maintain accurate data temporarily. CDSS have memory, which means that history of data can be obtained from the patient, and ventilator waveform data (VMD) is helpful to physicians as it can give information about a single patient that can be easily accessed via an app. Privacy issues are addressed in CDSS as it does not use private information of patients but instead treat them as anonymous subject.

CDSS integrate IoT and AI-based patient-monitoring devices, the analytics operating on real-time physiological data, and ML algorithms stand to improve diagnosis, prognostication, and adverse event recognition in the ICU. CDSS will reduce the mental burden on doctors and help in increasing quality of care and greater importance can be placed on every patient. This is helpful in effectively improving the battle against the coronavirus.

3.3.1 Wearable, Cuffless Blood-pressure Measuring Devices

Blood pressure (BP) is one of the important parameters that has to be continuously monitored for a patient who has COVID-19 symptoms. Research has shown that a person with high blood-pressure symptoms are more likely to get COVID-19 as they lead to a weak immune system. It is also important to monitor the BP levels of patients who are tested positive for COVID-19 and are either in home quarantine or in hospital. So, it is better to use a wearable device that measures blood pressure and is connected to the Internet as data to be monitored by doctors.

The objective of the standard P1708 is to establish a performance evaluation of wearable, cuffless BP measuring devices. This standard is not dependent on the type of device, nor to another device in which the BP measuring device is either attached or embedded. This is useful as it is applicable to almost all types of wearable BP measurement devices that have different modes of operation (e.g., to measure short-term, long-term, snapshot, continuous, beat(s)-to-beat(s) BP, or BP variability).

Different from conventional devices that mainly use an inflatable cuff accompanied by a pressure transducer and/or Korotkoff sounds detector, wearable cuffless devices utilize different kinds of sensors for collecting data and/or for achieving the calibration process without using an inflatable cuff. These sensors may include optical transmitter and detector, accelerator, pressure transducer, electrodes, capacitive sensors, organic phototransistors, an epidermal sensors based on flexible and printable electronics, and so forth [9].

Such devices allows communication between wearable cuffless devices and computing devices with the help to network connectivity. Thus, they can be integrated with an app and provide data to healthcare specialists. They, in turn, can give feedback to patients. This is all possible due to the networking part of this device.

3.4 Internet of Things Buttons for Real-time Notifications in Hospital

In hospitals, doctors and nurses perform particular tasks routinely and with the appearance of COVID-19, doctors-to-patient ratio is very high, so that good communication and time are not given. Hence, here comes the use of IoT buttons.

IoT buttons are types of IoT devices that help in performing certain actions if pressed. An illustration of IoT button is depicted in Figure 3.3. Restroom cleanliness is one of the important components in hospital quality. Due to their active usage, it can be difficult to detect the presence of dirty restrooms

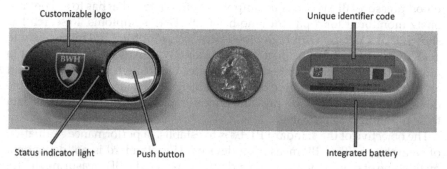

FIGURE 3.3
A picture of IOT button. [10]

that need to be cleaned. IoT buttons can be used by users to report restrooms that require cleaning by the staff and can request this when required, which helps in less direct contact as well as efficient communication [10]. It can also help in giving feedback by the user as well.

IoT buttons are small and unobtrusive devices that have been developed by a wide range of technologies. IoT buttons can be used for tasks like room cleaning, if required, or for more general tasks such as calling a nurse. IoT buttons can be customized to send a message embedded in the device through the Internet to anyone to whom the person wants the message to be sent. For example, since close relatives cannot be far away if a person is infected, these buttons will help in sending appropriate messages to the concerned person. The notification can generate a message, page, or an email. The entire process flow after the press of IoT button is indicated in Figure 3.4. In addition to these notifications, the device can also be programmed so that calls are made or reminders left for the person to take tablets through the hospital management system. The privacy issues can be diverted with the help of encryption or coded mode of communication where only the hospital staff can understand. An IoT button can perform many different actions through various types of button presses (single button press, double button press, and long press).

When the IoT button is pressed, a message is sent with the help of a wireless network to the main server. The server can be configured to send notification fit for the purpose. The notification can lead to many routines. The routine may be a text message or mail or may generate a page. In addition to this protocol, they can be configured in any manner using application programming interfaces (APIs), software methods that allow computer systems to exchange information.

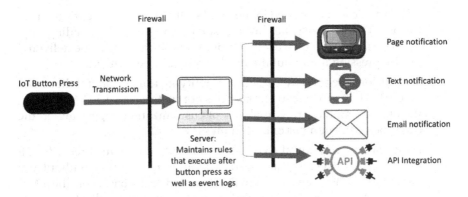

FIGURE 3.4
Schematic of the process flow of an IoT button. [10]

IoT buttons have a wide range of applicability in hospitals. They can be divided into two categories: patient and hospital staff applications, depending upon the main person who uses the IoT button.

IoT buttons might be a substitution for a call button for patients. The IoT button can be either be a mobile that has wireless access, or may be connected to a rail of a hospital bed and is programmed as a call for help. In addition to this, IoT button can be used for a further depth of information depending on the number of presses: for example, a patient will have different concerns to which the solution is depth of information. He may use a single button press to call for help, but a responding nurse could then provide a long press to indicate that the call has been addressed or provide two presses to request additional help. The ease of button use may improve the reliability and validity of patient-reported outcomes.

The applications of IoT buttons regarding hospital applications focus not only on streamlining notifications altogether, but real-time data can be gathered that will help in improving quality of workflow development. For example, an IoT button can be placed on a linen cart that will help housekeeping staff when the linen needs to be restocked. Recognizing the pattern of the restock requests, it can be analyzed. This data can be used for recommendation and justifying the changes as a response given by housekeeping staff. In addition, IoT buttons help in detecting improper working in workflow with patients. Button presses may also be used to flag in-patients who are ready for discharge after they have recovered from COVID-19 to the respective authority.

IoT buttons are very helpful, both to patients and doctors. These buttons require network connectivity, but other than this there is no disadvantage to these buttons. The advantages, on the other hand, are considerable.

Following are some of the advantages when compared to existing similar applications:

- QR Code Readers are used by hospital staff to know a patient's medical history. This is accessible through smartphone with QR reading app. When compared to this, using IoT button notifications can be delivered just by pressing it without the need of any user equipment.

- Call buttons are used by patients for notifying nursing staff if they need help. This has only one function and is not customizable. IoT buttons on the other hand have the capability of customization depending on the duration or the number of button presses.

- EMR (electronic medical records) are a standard by which records of a particular patient can be accessed and monitored digitally without any paperwork. They can also signal completed tasks based on the EMR changes. However, the drawback is that each new application needs programming to be done, which is time consuming. IoT buttons provide an easy and simpler interface for the modification of such tasks.

TABLE 3.1

Summary of IoT and AI Technologies Used for COVID-19

Type	Capability	Examples
Iot-Q-Band	1) Helps in tracking Quarantine subjects 2) Easy maintenance	Hong Kong electronic wristband Electronic ankle bracelet in USA
Smart Thermometers	1) Touch free and Wireless 2) Increasing the diagnosis rate	Kinsa,Tempdrop,iSense,iFiever
Smart Helmet	1) Temperature Capturing 2) Less Human interactions 3) Facial and location capture	KC N901 in China

3.5 IoT-based Smart Helmet for COVID-19

IoT Arduino, Node MCU, and thermal camera based settings mounted on the helmet is proposed in [11]. The system designed for the helmet includes two cameras, which try to do thermal imaging and facial recognition and classifies the object as a person or not. When the thermal camera identifies high body surface temperature, then it issues a warning via a mobile app and notifies the wearer. Even though this configuration might be costly, these systems can be deployed in densely populated areas in big cities such as Mumbai, Seoul, New York, and so forth, where daily interactions cannot be avoided – this idea helps in unavoidable close-contact situations. Table 3.1 compares IoT-Q-Band, Smart Thermometers and Smart Helmet in terms of their capability.

3.6 Sanitization Using IoT and AI Technology

Monitoring of COVID -19 patients in small-, as well as large-scale to curb the spread of the disease is discussed but, even with that, COVID-19 is spreading because the respiratory particles of the victims are spread onto surfaces and if we come into contact with it and it makes its way into our body, we too will be affected. One of the early cases of COVID-19 in Germany was through a handshake, showing how infectious this virus is. The solutions for the above stated problem are to use mass automated hand sanitizers that will help in curbing the spread as well as in maintaining good hygiene and health in the general public.

3.7 Ultraviolet Light Surface Disinfection Devices

Since the COVID-19 virus stays for 6 hours to 9 days on surfaces, other than sanitizing the surface there is no other way to remove the virus. UV light is a form of electromagnetic radiation with more energy than visible light, but less energy than X-rays. The higher-energy UV-C rays can damage DNA and RNA via of the viruses, thus preventing the replication of these microorganisms. The PX-UVC device (Xenex Disinfection Services) uses the Xe-gas flash lamp to generate high energy, ultraviolet, and visible light, in microsecond bursts (waveforms) at 67 per second frequency. No-touch UV technology is dependent on the distance between the lamp and the surface being disinfected. The inverse square law states that twice the distance between the lamp and the surface being disinfected will make it four times the time required for disinfection. The PX-UVC uses five-minute cycles to clean surfaces with the UV-C rays. In order to achieve optimal efficacy, high-touch surfaces are within two meters of the lamp surfaces. The number of cycles is configured taking into account how much it has to be sanitized. As UV-C spectrum electromagnetic waves can harm humans, it is done when the room is empty. As a precautionary motion, sensors are placed such that even if a person enters unknowingly, the machine stops and stays off until the person leaves [12].

Working of UV-Disinfectant:- The PX-UVC device reacts in a way such that ozone is produced, hence the room has to be well-ventilated .These robots do not leave any sort of residue and help in ideal cleaning of the surface .The robot allows access to the room (a green light turns on) after a delay that allows the ozone to dissipate. In [14], rooms were aerated after using the robot. These machines have large-scale as well as no-contact sanitation applications using Swarm technology, where each bot can communicate with each other and make a path plan and efficiently sanitize the place thoroughly. If the application of Swarm tech deems it too costly, we can always plant cameras and control it via a WiFi controller or a Bluetooth depending on the range. Through this concept, IoT and AI are used in efficient organization of the bot.

3.8 Drones and other Robots for Spraying Disinfectant

Drones are used in delivery system as well as thermal imaging over a certain area, and can also be used for spraying pesticides, germicide in the agricultural sector. IoDT (Internet of Drone Things) is used for spraying the disinfectant over selected regions and without human contact to help in cleaning the surfaces. There are pumps and a nozzle, and it sprays into the air as well as sprinkling it on the surface. For disinfectant spraying, the roads or an area can be selected on Google Maps. The drones can record videos and can also be used to make announcements before the spraying starts. There have been industry-related drones all around the world helping in fighting this virus. These automated disinfecting drones are capable of spraying disinfectants from 3 to 450 feet in height, covering long distances in a short time without involving public workers at the site of spray deployment. Marut Drones, a Hyderabad-based startup led by a team of IIT graduates, recently launched an entire line of drones to combat the COVID-19 pandemic in India. The company has drones for sanitizing, medicine delivery, thermal analysis, movement monitoring, and crowd surveillance in its arsenal to oppose COVID-19. These drones can be controlled via a smartphone-enabled remote controller or can be autonomous, using AI algorithms which helps in path planning and taking into account human activity and, if someone is in the vicinity of sprayed disinfectant, it will stop and then continue after the person clears the area. The above-mentioned robots are widely used in spraying alcoholic sanitizers over large areas when no human activity is happening. These robots are remotely controlled to avoid hazardous contact with the disinfectant spray [15]. With the application of Swarm technology these drones' future scope is to communicate with each other and deliver optimal solutions.

3.9 IoT-Enabled Smart City during COVID-19

Smart city is a concept of idealizing a city by connecting all the network-based elements. It means implementation of all information and communication technologies, connected in a network. This helps in effectively operating the city to combat COVID-19 pandemic. Quarantine is an important part in preventing the spread of coronavirus, but it is not always feasible for the authorities to monitor each and every part of the city. This is where the concept of smart city, which includes 5G, IoT and AI technologies are being incorporated.

Data is collected from smart sensors that are embedded in IoT networks and Artificial Intelligence is implemented in populated places like marketplace, airports, train stations can help to fight against both the present and future pandemics. Smart city infrastructure can help in social distancing norms by introducing transportation technologies like crowd monitoring, smart parking and traffic rerouting.

In smart cities the patients can be remotely monitored by doctors. IoT Telemedicine allows all conversations to be virtual to serve patients better.

This can be done with either a website chatbot or smartphone application. In many countries an AI-based bot has been implemented to help patients. This helps doctors avoid repeating the same information for each patient. Smart cities allow residents to implement smart living in their houses. For example, Smart Doorbells can be implemented so that users can be prevented from touching them [16].

In general, the efficiency of smart city varies on the connected device together, but it also has its privacy issues because of continuous data capture from multiple devices. When the pandemic is over, it is important to determine what has to be done with the data collected from the people. An example of this issue is Arogya Setu app, developed by government of India, that helps in combating COVID-19 as it tracks the travel history of a COVID-19 positive person and his place of living.

3.10 Conclusion and Future Scope

Different implementations of IoT and AI to combat this virus are discussed. It is understood that monitoring and screening, and proper diagnosis and planning by the government, can establish suitable techniques for combatting the virus. Lessons learnt from the COVID pandemic can be used to prepare for the future by implementation of smart city using techniques like IoT and AI.

- Though technologies used in the field of IoT and AI to combat the virus exist, they have not repressed it to a maximum extent, hence, faster network connection and faster algorithms are required to perform the action efficiently and at low costs to make them effective products.

- This pandemic is a reminder of how cleanliness is the key, so it is required to implement smart entry (screening and hand sanitization) in all public places, and long after COVID-19 is curbed, social distancing can be maintained with the help of IoT wearables and other products.

- Devices such as IoT buttons, drones and thermal cameras will be the new normal in the near future, as they are going to help us in case another deadly virus breaks out.

References

[1]   Sedov, L., Krasnochub, A., and Polishchuk, V., "Modeling quarantine during epidemics and mass-testing using drones", *PLoS ONE* 15(6): e0235307, 2020, https://doi.org/10.1371/journal.pone.023530.

[2] Punn, Narinder, Sonbhadra, Sanjay, and Agarwal, Sonali, "Monitoring COVID-19 social distancing with person detection and tracking via fine-tuned YOLO v3 and Deepsort techniques", 2020.

[3] Jiang, Zheng, et al. "Combining visible light and infrared imaging for efficient detection of respiratory infections such as COVID-19 on portable device." arXiv preprint arXiv:2004.06912, 2020.

[4] https://www.flir.in/discover/public-safety/faq-about-thermal-imaging-for-elevated-body-temperature-screening/

[5] Bai, L., Yang, D., Wang, X., Tong, L., Zhu, X., Bai, C., and Powell, C. A. (2020). Chinese experts' consensus on the Internet of Things-aided diagnosis and treatment of coronavirus disease 2019. *Clinical eHealth*. doi:10.1016/j.ceh.2020.03.001.

[6] www.xenex.com/resources/news/martin-health-system-unveils-xenex-germ-zapping-robot/.

[7] Singh, Vibhutesh, Chandna, Himanshu, Kumar, Ashish, Kumar, Sujeet, Upadhyay, Nidhi, and Utkarsh, Kumar, "IoT-Q-Band: A low cost internet of things based wearable band to detect and track absconding COVID-19 quarantine subjects", *EAI Endorsed Transactions on Internet of Things*, 2020, 20. 163997. 10.4108/eai.13-7-2018.163997.

[8] Rehm, Gregory B., et al. "Leveraging IoTs and machine learning for patient diagnosis and ventilation management in the intensive care unit". *IEEE pervasive computing* vol. 19, 3 (2020): 68–78. doi:10.1109/mprv.2020.2986767.

[9] Arakawa, Toshiya. "Recent Research and Developing Trends of Wearable Sensors for Detecting Blood Pressure". Sensors (Basel, Switzerland) vol. 18,9 2772. 23 Aug. 2018, doi:10.3390/s18092772.

[10] Chai, P. R., Zhang, H., Baugh, C.W., Jambaulikar, G.D., McCabe, J.C., Gorman. J.M., Boyer, E. W., Landman, An Internet of Things buttons for real-time notifications in hospital operations: Proposal for Hospital Implementation. *J Med Internet* Res, 2018, 20(8):e251.

[11] Abdulrazaq, Assoc. Prof. Dr. Mohammed, Zuhriyah, Halimatuz, Al-Zubaidi, Salah, Karim, Sairah, Ramli, Rusyaizila, and Yusuf, Eddy. (2020). Novel Covid-19 detection and diagnosis system using IOT based Smart Helmet. *International Journal of Psychosocial Rehabilitation*. 24. 2296–2303. 10.37200/IJPR/V24I7/PR270221.

[12] https://xenex.com/?utm_source=google&utm_medium=cpc&utm_campaign=brand

[13] Casini, B., Tuvo, B., Cristina, M. L., et al. Evaluation of an ultraviolet C (UVC) light-emitting device for disinfection of high touch surfaces in hospital critical areas. *Int J Environ Res Public Health*, 16(19): 3572, 2019; Sep 24. doi:10.3390/ijerph16193572

[14] Preethika, T.; Vaishnavi, P.; Agnishwar, J.; Padmanathan, K.; Umashankar, S.; Annapoorani, S.; Subash, M.; Aruloli, K. Artificial Intelligence and Drones to Combat COVID - 19. Preprints 2020, 2020060027 (doi: 10.20944/preprints202006.0027.v1).

[15] Nasajpour, M. et al. "Internet of Things for current COVID-19 and future pandemics: An exploratory study." ArXiv abs/2007.11147 (2020): n.p.

4

Potential Contributions of AI against COVID-19

Sreeja Sarasamma and Yashbir Singh

CONTENTS

4.1 Introduction

At the end of 2019, a novel strain named Severe Acute Respiratory Syndrome Coronaviru2 (SARS-Cov-2) caused the outbreak of COVID-19, a worldwide pandemic with catastrophic consequences for populations and healthcare systems across the globe. Currently, vaccines or drugs are available to treat covid-19, and a major number of deaths are reported in elderly patients. COVID-19 and its pandemic has disrupted society in a major way. Healthcare innovation and artificial intelligence are needed more than ever before as vital forces to meet the myriad challenges of pandemics. Researchers and developers are increasingly using artificial intelligence (AI), machine learning (ML), and natural language processing to track and contain coronavirus, as well as gain a more comprehensive understanding of the disease.

DOI: 10.1201/9781003126218-4

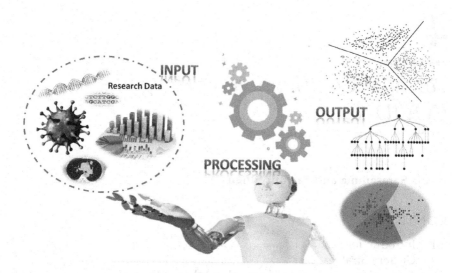

FIGURE 4.1

Figure 4.1 shows the possible applications of AI in the COVID-19 pandemic (Figure 4.1).

In this pandemic era, health-system capacity has rapidly increased in public and private systems which help in providing essential supplies and personal protective equipment (PPE) for health workers. We have focused on the development of rapid diagnostic kits and building makeshift hospitals. Also, for short-term containment, a new research push was required for the manufacturing of personal protective equipment. AI played an important role in these efforts. The characterization of material used for PPE and understanding of its properties will provide a foundation of the manufacturing of PPE using AI software [2,3, 4, 5]. On the other hand, we have to focus on recruitment of and training more healthcare staff. In the future, digital healthcare will play a vital role in overcoming the increasing demands of infected candidates and trace-case contacts – using the mass serological diagnosis to provide the serological ID that helps them to work, to protect healthcare workers, to travel, and to understand the trajectory of the transmission within the community. In the long term social distancing and masks are essential until vaccines finally reduce infections to decisive levels.

These things had made a difference very quickly. But we live in the digital world: we can use artificial intelligence methods that could be even more effective for society to overcome this pandemic. In this case, the drone delivery system could work for isolation cases, and the robot intelligence systems could work for the infected patients' care and the last one possible way, thermal imaging concept for pre-screening individuals with abnormal temperatures, which work in quite an effective way.

4.2 Current Strategy

4.2.1 Containment ("Epidemiological Avoidance")

Identification and tracking of infected individuals with isolation and quarantine at the start of an outbreak help us slow the spread of COVID-19. This can be achieved by contact tracing, which involves identifying all of the people that patient has contacted; inform those individuals that they may be at risk and then testing or isolate them. Asian countries like South Korea and Taiwan had a successful containment – tested on a massive scale, people coordinated governmental response, and prompt contact testing and quarantine [1].

4.2.2 Testing Is Critical

Massive testing is essential to ensure that people living and working in high-risk settings have access to FDA-approved tests. These tests would be easy to perform, readily available outside of the healthcare setting, quick with results and inexpensive to people. However, testing alone is insufficient but vital to any kind of screening program is the action taken once the test result is positive.

In general ways, there are few strategies used by the people, such as personal hygiene measures, social distancing high-risk individuals, bans on crowds, and lockdown of the infected region. Apart from that, our clinicians and medical engineers have been involved in this fight against with COVID-19.

4.3 Role of AI: From Diagnosis to Outcome Predictions

Recently, many clinical platforms of AI to the COVID-19 pandemic response have focused on diagnosis with medical imaging, with multiple numbers of studies focused on non-invasive monitoring approaches. There have been numerous scientific blogs and articles on the fight against COVID-19 using AI. Here, we present a few enabled AI strategies:

4.3.1 Isolation: Drone Delivery

COVID-19 is making us rethink the possible path we usually follow. Being social, we like to have social meetings but, this pandemic era is making us consider social distancing. In the current situation where the virus spreads very fast through touching, being physically isolated is the one way to overcome this disease in the absence of COVID-19 vaccines. In this case, drone delivery service of the very important items is an initiation for societies during this time of maintaining social distancing. Accepting the impact of drone delivery in this era of health emergency, governments are encouraging many more drone companies to develop their capacity to serve humanity in various ways [6, 7].

4.3.2   Equipment: 3D Printing

As we can see, healthcare systems are completely overloaded and fatigued by COVID-19. Scientists and designers have volunteered their relevant skills to ease the pressure on supply chains. 3D printing industries are now more visible, making 3D printed face shields and 3D printed respirator valves. While, in the novel spirit and creativeness of the 3D printing societies is commendable, medical devices are complicated and reverse engineering can have surprising outcomes [8].

4.3.3   Care: Intelligent Robot

This pandemic time has and it is time to speak technologically. Artificial intelligence and robots are increasing, perhaps one day to replace human workers entirely. It would be amazing if we could protect our doctors and nurses by turning more work over to robots. Moreover, robots could facilitate a form of telemedicine that would keep individuals out of areas of infection. COVID-19 could play a very important role, like a catalyst, in developing intelligence robotic systems that can be speedily deployed, for remote access, by the experts and important service suppliers without travelling. It is not very difficult to imagine a future in which a delivery robot brings food and materials to isolated individual's homes, avoiding potentially infecting them [9].

4.3.4   Data: Internet of Things (IoT)

The goal of the IoT is to prevent the spreading of coronavirus, which can be transmitted by touch and by air as well on body surfaces. This is exciting to this IOT platform to improve the air quality in societies. There are especially few possible technologies so far that works behind the IoT technology from improving the society's air quality [10, 11].

- Inexpensive IoT sensors can be distributed throughout societies that could monitor the air quality in the local areas.
- A cloud-based AI platform measures the data that is collected from sensors and can evaluate air quality.

In a similar fashion, Zoom, Face Time, and Skype are becoming very popular because they connect the people to remotely communicate with machines for virtual inspections and remote diagnostic support.

4.3.5 Model: Deep Reinforcement Learning

This is an artificial intelligence approach that explores the environment and receives input based on rewards, which is defined already in the environment. This approach could work more closely in the battle against COVID-19's spread. In this harsh time that everyone knows, we need to learn from the environment how to take effective action against this disease. In this kind of study, we have to understand the costs ratio of our decisions, considering a multitude of factors like local and global economic impact, public health, and psychological effects. This type of study addresses reinforcement learning, especially integrating it with deep neural networks known as deep reinforcement learning [12, 13].

4.3.6 Drugs: Generative Design Algorithms

Machine learning can help to advance drug-development process, providing a deep understanding of current antiviral benefits, forecasting infection rates and, in the faster patients, screening. There are various other suitable application areas, but the problem is with limited data and the ability to combine composite structures into deep learning models. A majority of convolutional neural networks techniques are shown that are connected to molecular interactions.

- Drug screening using the deep learning approach for COVID-19 by (Zhang, et al.) has used current antiviral drugs that might help to COVID-19 patients. This work is being suspended so that Adenosine and Vidabrine compounds might potentially work.
- In another other research, they used AlphaFold library to find the protein structures of Covid-19 virus. AlphaFold is a deep learning library for computational chemistry. With these protein structures (if correct), scientists will gain close access to the molecular structure of the virus, which could potentially pave the way for finding vaccines.

4.3.7 Radiology: CT Modalities

Chen et al. research has developed deep learning method for the detecting [of] COVID-19 in CT modalities and [the result is shown as very good, specific and sensitive]. We could say, CT scans potentially will enable a better treatment. This deep learning concept on imaging could help minimize the burden. Moreover, learning how disease manifests itself in CT-scans could help present more closely into the disease itself [14,15].

In conclusion, these are potentially quite impactful approaches of artificial intelligence to fighting against the Covid-19 though, mostly they are still in the early stage. However, limited data sharing continues to inhibit the entire progress in medical research problems. We believe that utilizing things like meta-learning, domain adaptation, and reinforcement learning, while loosening restrictions to healthcare data, could allow machine learning to play a very significant role in containing/responding to both COVID-19 and future pandemics.

References

[1] Shaw, R., et al. (2020). "Governance, technology and citizen behavior in pandemic: Lessons from COVID-19 in East Asia." *Progress in Disaster Science*: 100090.

[2] Ultraporous nanofiber mats and uses thereof. *UCN104936671A, JP2017200940A US14648925, EP2928577A1, EP2928577A4, WO2014093345A1, Patents.*

[3] Patel, J. P., Zou, G., Hsu, S. L, et al. (2015). Path to achieving molecular dispersion in a dense reactive mixture. *Journal of Polymer Science Part B: Polymer Physics*, 53, 1519–1526.

[4] Patel, J. P., Hsu, S. L, et al. (2016). An analysis of the role of non-reactive plasticizers in the crosslinking reactions of phenolic resins. *Journal of Polymer Science Part B: Polymer Physics*, 55, 206–213.

[5] Patel, J. P., Hsu, S. L, et al. (2016). An analysis of the role of reactive plasticizers in the crosslinking reactions of phenolic resins. *Polymer*, 107, 12–18.

[6] Passi, G. R. (2020). Novel coronavirus (COVID-19) epidemic. *Scientific American*.

[7] Skorup, B., and Haaland, C. (2020). How drones can help fight the coronavirus. Mercatus Center Research Paper Series, Special Edition Policy Brief (2020).

[8] Ishack, S., and Lipner, S. R. (2020). Applications of 3D printing technology to address COVID-19 related supply shortages. *The American Journal of Medicine*.

[9] Bullock, J., Pham, K. H., Lam, C. S. N., and Luengo-Oroz, M. (2020). Mapping the landscape of artificial intelligence applications against COVID-19. arXiv preprint arXiv:2003.11336.

[10] Chen, M., Yang, J., Hu, L., Hossain, M. S., and Muhammad, G. (2018). Urban healthcare big data system based on crowdsourced and cloud-based air quality indicators. *IEEE Communications Magazine*, 56(11), 14–20.

[11] Dutta, J., Chowdhury, C., Roy, S., Middya, A. I., and Gazi, F. (2017). Towards smart city: sensing air quality in city based on opportunistic crowd-sensing. In proceedings of the 18th international conference on distributed computing and networking. January, pp. 1–6.

[12] Singh, Y., Wu, S. Y., Friebe, M., Tavares, J. M. R., and Hu, W. (2018). Cardiac electrophysiology studies based on image and machine learning.

[13] Arel, I., Rose, D. C., and Karnowski, T. P. (2010). Deep machine learning – a new frontier in artificial intelligence research. *IEEE Computational Intelligence Magazine*, 5(4), 13–18.

[14] Singh, Y., Shakyawar, D., and Hu, W. (2020). An automated method for detecting the scar tissue in the left ventricular endocardial wall using Deep Learning approach. Current Medical Imaging, 16(3), 206–213.

[15] Singh, Y., Shakyawar, D., and Hu, W. (2020). Non-ischemic endocardial scar geometric remodeling toward topological machine learning. Proceedings of the Institution of Mechanical Engineers, Part H: *Journal of Engineering in Medicine*, 234(9), 1029–1035.

5

A Comparative Study of COVID-19 Data Analysis Using R Programming

Yagyanath Rimal, Bharatendra Rai, Vijay Singh Rathore, Sakuntala Pageni, and Prakash Gautam

CONTENTS

5.1 Introduction

Coronavirus has a large variety of its origins, many authors argue, which may be transmitted from animal species, such as bats and pangolins, to the human population where it spread to human by air transmission and has become a pandemic [1] [2] [3] [4]. The most common transmission process is from droplets from person to person in close quarters when people are sneezing and by other communication means [5] [6]. According to authors [7] [8], the ongoing transmission is due to international travelers worldwide, possibly originating from Wuhan China [8] [9]. Johns Hopkins University and the World Health Organization (WHO) named the diseases caused by the virus, COVID-19 [10] [11]. The traveler's history searching and treatment with personal equipment were highly recommended for treatment using isolation precautions, for not community spreads were highly recommended yet [12] [13] [14]. Coronavirus disease is named after the deadly diseases SARS Cov 2. Coronaviruses have spiked S protein crowns [15] [16]. Inside the virus are genetic materials, RNA combined with a gene that has the capability of making more copies of the virus within the brief time when it is transmitted to living cells. The virus infects a person through the mouth, nose, and eyes, then reaches the lungs and the DNA releases its gene to promote the spread [17] [18] [19]. Then body cells pull the genetic materials from the virus to the body cells inside the air sack of the lungs. In human lungs are 6 million air sacs (alveoli) arranged throughout the upper, middle, and lower lobes. When blood comes into them it is oxygenated and passes through [17] [20] [21]. When a virus infects human lungs, it infects the entire lungs, blocking the oxygenation process of air sacs. This process leaks the blood inside the air sacs and further blocks the passing of oxygen to the blood capillaries around the lung's sacs. This process leads to the need for a ventilator for breathing. That ultimately leads to nonfunctioning air sacs in oxygenating the blood cells and evacuation of CO_2 gases from the lungs. The ventilator and prone position of patients when sleeping can help decrease the death rate with intensive hospital care [20] [22]. Once the disease reaches into air sacs it again releases generic materials, ribosomes, and generates more copies that infect more cells and cause a more acute condition that ultimately causes death. This process completes 2 to 14 days after symptoms first appear. Some notable symptoms are fever and shortness of breath observed in 20 percent of patients, while the other 80 percent has no serious symptoms [22]. Other symptoms include tiredness, body aches, stuffy nose, sore throat, vomiting, loss of appetite, loss of smell, and so forth. Most people have mild illnesses and can recover at home. According to a report [23], older people have a higher risk for cases of serious illnesses such as lung and heart diseases, diabetes, severe obesity, chronic kidney disease, liver disease, and asthma. High-risk people have a weaker immune system than the normal patient [24]. Even if you are not high-risk, social distancing of 2 meters (6 feet) is recommended, which helps prevent transmission of airborne virus person to person.

The testing process collects samples from lungs or oropharyngeal swabs and is stored in the viral transport medium. The standard method of testing the coronavirus is PCR (Polymerase Chain Reaction) used in the molecular biology of DNA replication [25]. Another test is for the coronavirus single RNA gene using reverse transcriptase-polymerase chain reaction procedures (RTPCT). The collected sample is mixed with linctus buffers for isolation of viral RNA from the sample and then isolation of virus DNA will be detected in positive cases.

Coronavirus (COV) causes illness ranging from the common cold to more serious diseases such as Middle East Respiratory Syndrome (MERS-Cov, 2012) and Severe Acute Respiratory Syndrome (SARS-COV, 2002) [26]. Coronavirus is zoonotic, that is, it can transmit between animals and human beings [26]. The primary source of coronavirus is believed to be a wet market in Wuhan, in December 2019, where both dead and live animals, including fish and birds, were sold [27] [28]. The Chinese authorities immediately tried to control the disease through many procedures such as placing patients in isolation to prevent airborne transmission of community spreading. COVID-19 causes more deaths during the initial disease in patients with asthma and/

or heart disease, who have less immunity power due to previous diseases [29] [30].

COVID-19 was first declared by Dr. Li Wenliang, a frontline health care worker who lost his own life fighting the disease in China [32]. The novel coronavirus became a public health emergency of international concern when it spread in many countries, killing and infecting people worldwide. On March 15, 2020, WHO declared the pandemic when there were five thousand dead and there were 10,000 people infected globally [31]. The countries most affected by the disease initially were China, Italy, Iran, and South Korea, where there was a high recovery rate thanks to sound hospital treatment and public awareness. However, the virus mutated and changed. It creates fever, common cold-like symptoms.

It was an almost universal convention to keep a distance from one another and wash hands with soap – the most prescribed measures against COVID. The habits of touching mouth and nose are always prohibited because of quick viral transmission. The mask is a core blocker for spreading as well as blocking the virus directly through the mouth and nose. Nurses and doctors always use personal protective equipment (PPE) with masks being compulsory while treating patients in the hospital.

It is supposed that there were 3,000 droplets spread from single cough by COVID patients. Virus spread by air can live three hours in normal conditions [31]. The steel and metal materials such as plastic will keep virus alive for more than three days. When the protective measures fail, COVID-19 diseases then spike protein on the virus histamine H2 receptors to cells of human lungs. [15] When the virus is being replicated inside the lungs, at first the human does not experience symptoms; this is called the incubation period, which lasts from 2 to 7 days, which ultimately increases symptoms of cough, fever, shortness of breath, dry cough, headache, sore throat, and pneumonia [34] [35].

The pre-symptoms are very abnormal in person-to-person spreading, so reporting cases and testing are significant for saving lives. The COVID-19 virus has RNA genetic materials whereas human genetic materials are DNA [36]. Covid-19 has positive single-strain RNA materials that could easily create protein and produce automatic RNA without body cells inside the human body. When protein and RNA create new copies, the illness easily could change its protein [37]. The lungs' agios receptors used in the blood purification process is highly affected by the human lungs combine and released virus RNA spread into phytoplasma. The ribosome and phytoplasma then create proteins and combines and creates viruses inside the lungs with making killing cell membranes of the lungs causes serious. Therefore, the metabolism activities of viruses can interrupt at any places. It is essential to find the presence of viruses inside the human body; however, the virus cannot be seen by the unaided eye, and the testing is not 100 percent accurate. The RTPCR test is for research purposes, whereas the rapid test is for large coverage with speed and lower cost.

The virus with single strain RNA is converted into double standard DNA with multiple copies and matched with sequencing DNA (ACGT) sequencing, whereas RNA uses AUGC sequences. There was a large company that developed the different kits for testing by sampling 10 to 20 special sequences that should be present in viral RNA. When RNA samples are first converted into a full sequence of DNA, therefore, this test requires more time to test. This process is called *reversed* because generally from DNA and RNA is generated to detect the cases. First, the complimentary copy of RNA (DNA) with florescent amplifying indicates the sample is considered positive when the florescent, not increasing, becomes negative [38] [39]. Whereas the rapid test uses the blood sample with a buffered solution, which indicates the antibody of human blood. The controller indicates the sample limits of the immunity system with Igm and Igg. These two were different only infected many times or first-time produced antibodies to the human body for the prevention of COV1 and COV2 diseases. This test requires waiting 10 to 15 minutes. The red line just indicates IGM implies the positive first time needs to be quarantine similarly when the two red lines appear in the kit indicates already been in positive therefore your blood produces antibody for treating COVID yourself and needs more precautions too [40] [39] [41] [42]. Whereas, if there are no red lines indicates you do not have the COVID disease confirmed. If the rapid test becomes positive or negative it is better to use the RTPCR test for confirmation.

5.2 Objective

The primary objective of this analysis is to explore the global trend of COVID cases using R Programming. The global trends and its patterns are analyzed using a comparative study of Nepal, India, and the United States as example cases for the data downloaded from Johns Hopkins repository (2020–21). Similarly, the specific objective of this study is to compare among developed, developing and underdeveloped country situation of COVID cases.

5.3 Methods

The focus of this research was to predict COVID cases using comparative approaches. The exploratory analysis included fourfold data preparation of COVID cases for the prediction of the next three months; data gathering, data preprocessing, applying the data mining algorithm, and analyzing the

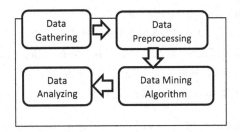

FIGURE 5.1
Research Method.

result using comparative as shown in Figure 5.1 constitute the key newness of this research. The data are collected from an open data source of the CSSEGIS repository, a world data repository maintained by Johns Hopkins University to r console after installation of "covid19. analytics" packages in R Programming. Then the data are classified, described according to their attributes before applying prediction and model design. Each line of coding is sufficiently explained to reach a conclusion.

5.4 Results

Here, researchers use the COVID data sets time series freely available on Johns Hopkins University repository from 01/22/2020 to 2020/09/12, from the reliable internet repository of https://raw.githubusercontent.com/CSS EGISandData/COVID-19/master/csse_covid_19_data/csse_covid_19_da ily_reports/09-10-2020.csv, retrieved on 2020-09-12 15:23:38 and stored in the variable as agg=covid19.data(case = 'aggregated') whose record structure follows. This is the World Health Organization's daily updated records repository, which stores data automatically downloaded, and at this writing has records of 3,954 cases of 14 variable attributes.

$ FIPS: int 45001 22001 51001 16001 19001 21001 29001 40001 8001 16003 ...

$ Admin2: Factor w/ 1788 levels "","Abbeville", 2 3 4 5 6 6 6 6 7 7 ...

$ Province_state: Factor w/ 549 levels "","Abruzzo", "Acre", 456 263 521

$ Country_regions: Factor w/ 188 levels "Afghanistan", 180 180 180

$ Last Update: Factor w/ 1 level "2020-06-23 04:33:22": 1 1 1 1 1 1

$ Lat: num 34.2 30.3 37.8 43.5 41.3 ...

$ Long_: num -82.5 -92.4 -75.6 -116.2 -94.5 ...

$ Confirmed: int 87 669 1032 1307 12 105 86 108 3932 10 ...

$ Deaths: int 0 34 14 22 0 19 0 4 153 0 ...

$ Recovered: int 0 0 0 0 0 0 0 0 0 0 ...

$ Active: int 87 635 1018 1285 12 86 86 104 3779 10 ...

$ Combined Key: Factor w/ 3756 levels "Abbeville, South Carolina, US"

$ Incidence Rate: num 355 1078 3193 271 168 ...

$ Case Fatality Ratio: num 0 5.08 1.36 1.68 0 ...

The tabular record looks like as and stores 3,798*14=3812 records.

Similarly, the all-time series records of confirmed case records could be easily loaded to r console and stored in variables. While storing data of each separate case of all and only confirmed could be loaded as using tsa=covid19. data (case='ts-All') and tsc=covid19.data (case = 'ts-confirmed') commands in r console, whose data structure in tabular format as stored records of (255*266) days records and total all-time series download r console 785 observation of 238 time-series data of COVID information.

The report: Summary (Nentries = 12, graphical. output = T) produces three outputs describing the total of 188 countries with 82 geographical locations of each confirmed cases worldwide. The entries equal to 12 produce the topmost 12 countries with increasing and decreasing trend patterns up to the date downloaded, where the researcher could easily count as is required. The output TS-CONFIRMED cases data dated as of 2020-06-22: 2020-09-11 19:33:22 includes the number of countries/regions reported, which is 188 while the number of cities/provinces reported is 82 locations. The unique

TABLE 5.1

Covid Aggregate Data Sets from Johns Hopkins

S. N	Province state	Country region	Last Update	Lat	Long	Confirmed	Deaths	Recovery
1	South Carolina	US	2020-07-05 04:33:46	34.2	–82.5	119	0	0
....								
3798		Zimbabwe	2020-09-12 15:23:38	–19.0	29.1	698	8	181

TABLE 5.2

Total Confirmed Cases Data Set from Johns Hopkins

Province_state	Country regions	Lat	Long	2020-01-22	2020-01-23	--	2020-09-12
1	Afghanistan	33.93911	67.709953	0	0		38572
....							
265	Zimbabwe	–19.015438	9.154857	0	0		7453

TABLE 5.3

Total Confirmed Cases Confirmed Summary

Country	Totals	%	Last	t-1	t-2	t-3	t-4	t-30
US	6396100	22.71	35888	33203	26387	50502	46156	56203
India	4562414	16.20	96551	95735	89706	86432	76472	66999
Brazil	4238446	15.05	40557	35816	14279	50163	43412	55155
Russia	1042836	3.70	5310	5172	5020	5064	4758	5054
Peru	702776	2.50	6586	4615	1598	13016	8619	0
Colom	694664	2.47	7813	15318	-315	8489	8497	12066
Mexico	652364	2.32	5043	4461	5351	6196	5824	5858
South A	644438	2.29	2007	1990	1079	2063	1846	2810
Spain	554143	1.97	10764	8866	8964	10476	9779	3172
Argentina	524198	1.86	11905	12259	12027	10684	11717	7663
Chile	428669	1.52	1642	1486	1267	1968	1870	1552
Iran	395488	1.40	2063	2313	2302	2026	2115	2510

number of distinct geographical locations of combined cases is 563. And the worldwide ts-confirmed totals are 28161434 cases.

The output further describes the global average percentage is 0.38 with a standard deviation of dispersion of 1.85 worldwide automatically calculated, whereas the top countries global percentage average many times different 7.02 percentage and the standard deviation is 7.37 implies the highest growth in top-scored countries. Similarly, TS-DEATHS cases data dated from 2020-06-22 to 2020-09-10 00:30:27 comprises 188 countries/regions of cities and provinces of 82 having 266 distinct geographical locations automatically generated. The worldwide ts-deaths reach is 909479 indicates its very increasing. This process further outputs of the top 12 countries with total its percentage and each day differences of one month's records as below.

Similarly, the records of the TS-RECOVERED cases from 2020-01-22 to 2020-09-12 15:23:38 19:42:25 explain the total of 188 countries of distinct geographical locations combined is 253. The worldwide ts-recovered reached is 18992383, and the current monthly growth rate of each top country displays.

Similarly, the AGGREGATED data ORDERED BY CONFIRMED cases from 2020-09-11 to 2020-09-11 19:42:25 of 188 countries of 549 providence of distinct geographical locations combined is 3,954 areas.

From the above table, the aggregate total cases with a percentage of confirmed, deaths, recovery, and active cases were displayed. Brazil has the highest (2.77) percent of confirmed case, where Lima has at least 1.45 percent among the top 12 countries. Similarly, the UK, Iran, Chile, and Russia highest lower confirmed case percentage. Similarly, UK (17.17), United States (10.69), and France (15.25) have the highest death rates, whereas South Africa and Saudi Arabia have 0.90 percent. However, the recovery rate in the United States, UK, and Lima has no recovery to have the least recovery rate where

TABLE 5.4

Total Deaths

Country	Totals	Per	Change	t-1	t-2	t-3	t-4	t-30
US	191766	3	907	1206	445	965	971	1505
Brazil	129522	3.06	983	1075	504	888	855	1175
India	76271	1.67	1209	1172	1115	1089	1021	942
Mexico	69649	10.68	600	565	703	522	552	737
United	41608	11.62	14	8	32	10	9	20
Italy	35587	12.57	10	14	10	11	9	10
France	30656	8.23	13	30	38	17	19	46
Peru	30236	4.3	113	147	138	337	153	0
Spain	29699	5.36	71	34	78	184	15	-2
Iran	22798	5.76	129	127	132	118	112	188
Colom	22275	3.21	222	442	-4	270	299	362
Russia	18207	1.75	127	141	121	119	108	128

TABLE 5.5

Total Recovered Cases

Country	Total	Change	t-1	t-3	t-4	t-30
Brazil	3657701	46069	39211	34843	35937	56890
India	3542663	70880	72939	70072	65050	56383
US	2403511	16032	28368	16497	17041	38800
Russia	859961	5892	6323	5749	5869	7104
South Africa	573003	3068	2206	2931	2597	5904
Colombia	569479	16594	23868	9070	11827	9358
Mexico	541804	3290	3373	4037	4238	3904
Peru	536959	0	14708	18346	7785	12534
Chile	401356	1801	1825	1839	1530	1878
Argentina	390098	7608	15900	9160	5707	5894
Iran	340842	1731	1697	1713	1632	1624
Saudi Arabi	299998	1032	720	1099	1148	2151

Chile and Iran have the best recovery on that day concerning yesterday's records. However, the United States, UK, and Lima have the highest active case percentage rate among top cases, whereas Chile and Turkey have a 10 percent active cases. The total plot describes similar output in the pie and bars diagram of deaths for visual display the records explain confirmed cases and aggregate and recovered of the top 12 countries. Here, the researcher just displays top 12 countries' information as sample cases. This could be only being selected 3 or all 210 countries. The first plot describes the bar plot of confirmed, deaths, and recovery of 12 countries. Similarly, the second plots describe each country's confirmed cases, deaths, and recovery cases. The total per location may use single or multiple countries to describe the linear regression (lM) by specifying its variables. The United States and Brazil have

TABLE 5.6

Aggregate Data Sets When Summarized

Location	Confirmed	%	Death	Recovered	%	Active	%
Sao Paulo, Brazil	312530	2.77	15996	162851	52.11	133683	42.77
Iran	237878	2.11	11408	198949	83.63	27521	11.57
Metropolitan, Chi	226128	2.01	5251	204387	90.39	16490	7.29
Moscow, Russia	224210	1.99	3929	156459	69.78	63822	28.47
Saudi Arabia	205929	1.83	1858	143256	69.57	60815	29.53
Turkey	204610	1.82	5206	179492	87.72	19912	9.73
Maharashtra, India	200064	1.78	8671	108082	54.02	83311	41.64
France	195546	1.74	29812	72092	36.87	93642	47.89
South Africa	187977	1.67	3026	91227	48.53	93724	49.86
Lim	163824	1.45	4814	0	0	159010	97.06

the highest number of cases, but Germany and Italy have the lowest number among them. Similarly, the total per location could easily use confirmed data with geolocation of country comparisons between many country-like tots. per. location (tsc, geo.loc=c ('India', 'US')) produce as the output of Linear Regression (lm) and GLM using family of INDIA with 4562414 cases.

Coefficients:

Estimate Std. Error t value Pr(>|t|)			
(Intercept)	-899460.3	92692.5 -9.704	<2e-16 ***
x.var	13984.0	686.8 20.360	<2e-16 ***

Residual standard error: 705200 on 231 degrees of freedom
Multiple R-squared: 0.6422,
Adjusted R-squared: 0.6406
F-statistic: 414.5 on 1 and 231 DF, p-value: < 2.2e-16

The India linear regression formula, which produces five-level summaries of minimum maximum, median, 1Q, and 1Q, are -803619 -626998 -136732 479944 2203610, respectively. The coefficients output explains that the model intercept describes the negative association (-899460.3) when x-intercepts at the y-axis. The other intercepts indicate individual associations when x is zero for predicting the explanatory variable is 13984.0. Similarly, the standard error is 686.8, which indicates the variability error of sampling data is 68 percent. Similarly, the t value implies 20.360 is the ratio between standard error and estimation is mostly concerned with the mean of confirmed cases of COVID cases of India. Similarly, the last output p-value (<2e-16) indicates its model and other explicatory variables' significance. This evaluates with 0.05 for rejection or acceptance. The three stars indicate these two variables have highly significant to one another. The root means the square error is 705200 implies how well is the model for prediction model, indicates in every 231

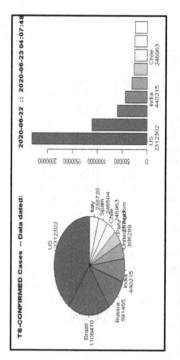

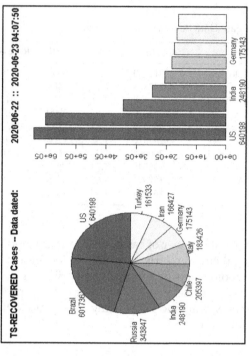

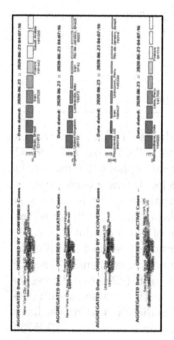

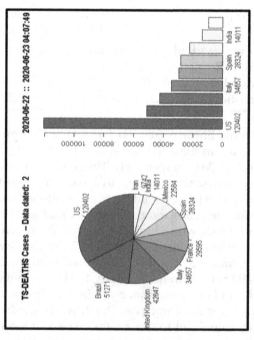

FIGURE 5.2

Total Plots of Confirmed, Deaths and Aggregate of Top Countries.

data not accurately degrees of freedom. The multiple r square is 0.6422implies a percentage in the variation of x explained by the explanatory variable. Likewise, the adjusted r-squared is 0.6406 is the ratio between estimation and error terms in the model. The f-statistic is 414.5 indicates the model significance on 1 by 99 DF. The p-value is less than 2.2e-16 implies a rejection of the null hypothesis, and these two variables are significant associations.

Similarly, the United States with 6396100 cases Linear Regression (lm) and GLM model output produces implies the there is a significant association between them

Min	1Q Median	3Q	Max
-801.2 -573.0	84.7	260.9	405.4

Coefficients:

	Estimate	Std. Error	z value	Pr(>|z|)
(Intercept)	1.176e+01	1.829e-04	64277	<2e-16 ***
x.var	1.805e-02	9.746e-07	18518	<2e-16 ***

Null deviance: 530700513 on 232 degrees of freedom
Residual deviance: 45592008 on 231 degrees of freedom
AIC: 45595166 with Number of Fisher Scoring iterations: 25

At this moment the US cases whose AIC is 45595166 when the number of fishers scoring iterations at 5 indicates another model design. From using this model could easily change any country information.

The above two charts, describe 266 days' records (from 2020/01/22 to 2020-09-12 15:23:38) produce in the same chart which clearly describes Covid's increasing trend. When the logs chart is in curve with concave in the first plot indicates its decreasing trend of cases whereas the concave pattern indicates a decreasing trend. The red line indicates the total death rate is quite smaller in India than in the United States.

These plots display the total growth rate with comparison among Nepal, India, and the United States with its number in bars of total cases. The total growth rate of confirmed cases is low in Nepal due to the lack of proper testing facility could not be mentioned by the government in the early days. The United States has the highest because it focuses on testing more cases for the virus spreading control infection the multiple country diagram with total cases could compare with growth using growth Rate (tsc, geo.loc =c('Nepal', 'India', 'US')) command plots.

The total plots directly provide the country present status with a line chart describes with a comparison of Nepal India and the United States each selected country's current status, confirmed cases, death on each day since the beginning. The green line describes each day death cases from the beginning. the combined plots and individual country plots describe increasing

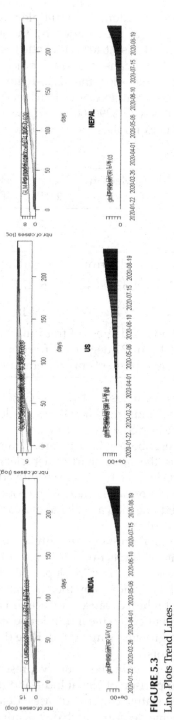

FIGURE 5.3
Line Plots Trend Lines.

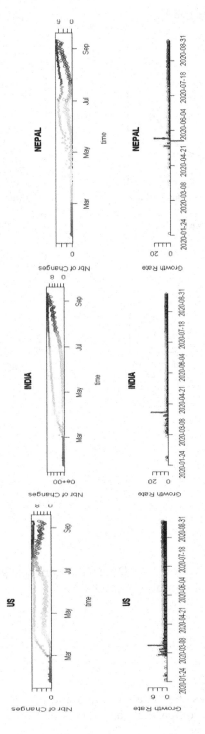

FIGURE 5.4
Growth Rate Plot.

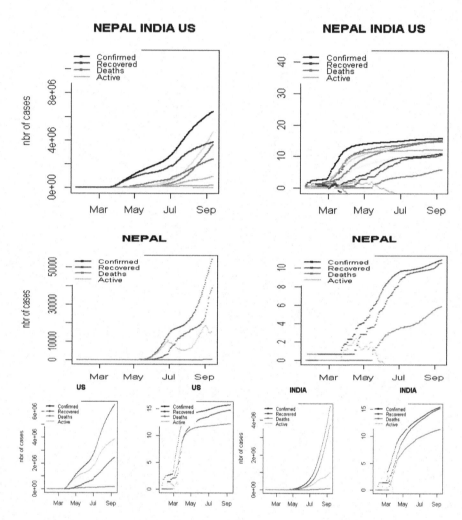

FIGURE 5.5
Total Line Plots.

trends. Therefore, control of COVID needs urgent using holistic measures from authorities control community spread.

5.5   Conclusion

The COVID-19 diseases started from China and spread all over the world to 210 countries. This comparative analysis is based on the data taken on

2020-09-12 15:23:38 from the Hopkins repository. Brazil has the highest number of cases per capita followed by India, and the United States has the highest number of cases. However, current trends are still increasing worldwide. Each government should take appropriate measured for better prevention. The United States has the highest death rate than India and Nepal however it's in increasing trend is quite flatten than India a developing and underdeveloped country Nepal.

5.6 Key Messages

COVID-19 a viral disease spread around world from China, first recorded in January 2020, that quickly covered the world in a pandemic. The data from countries the authorities are being regularly maintained at Johns Hopkins University repository. Its comparison growth rates, patterns, and lessons learned from those countries are necessary for controlling its community spread. Therefore, this chapter compared three sample countries – developed (the United States), developing (India), and underdeveloped (Nepal) – using R Programming.

References

[1] R. Woelfel, V. Corman, W. Guggemos, and M. Seilmaier, "Clinical presentation and virological assessment of hospitalized cases of coronavirus disease 2019 in a travel-associated transmission cluster," 2020.

[2] A. Rodriguez-Morales and D. Bonilla-Aldana, "COVID-19, an emerging coronavirus infection: current scenario and recent developments-an overview," 2020.

[3] I. Chakraborty and P. Maity, "COVID-19 outbreak: Migration, effects on society, global environment and prevention," 2020.

[4] C. Contini, D. Nuzzo, N. Barp, and A. Bonazza, "The novel zoonotic COVID-19 pandemic: An expected global health concern," 2020.

[5] R. West, S. Michie, G. Rubin, and R. Amlôt, "Applying principles of behaviour change to reduce SARS-CoV-2 transmission," 2020.

[6] N. Zhang, B. Su, P. Chan, T. Miao, and P. Wang, "Infection spread and high-resolution detection of close contact behaviors," 2020.

[7] B. Neupane and B. Giri, "Current Understanding on the Effectiveness of Face Masks and Respirators to Prevent the Spread of Respiratory Viruses," 2020.

[8] Y. L. T. R. A. K. K. Prem, "The effect of control strategies to reduce social mixing on outcomes of the COVID-19 epidemic in Wuhan, China: A modelling study," 2020.

[9] M. Cascella, M. Rajnik, A. Cuomo, and S. Dulebohn, "Features, evaluation and treatment coronavirus (COVID-19)," 2020.

[10] P. Christidis and A. Christodoulou, "The Predictive Capacity of Air Travel Patterns during the Global Spread of the COVID-19 Pandemic: Risk, Uncertainty and Randomness," 2020.

[11] S. Lauer, K. Grantz, Q. Bi, and F. Jones, "The incubation period of coronavirus disease, 2019." (COVID-19) from publicly reported confirmed cases: estimation and application, 2020.

[12] A. Ferhani, "The International Health Regulations, COVID-19, and bordering practices: Who gets in, what gets out, and who gets rescued?" 2020.

[13] E. Torri, L. Sbrogiò, R. Di ,and S. Cinquetti, "Italian Public Health Response to the COVID-19 Pandemic: Case Report from the Field, Insights and Challenges for the Department of Prevention," 2020.

[14] B. Coutard, C. Valle, X. Lamballerie, and B. Canard, "The spike glycoprotein of the new coronavirus 2019-nCoV contains a furin-like cleavage site absent in CoV of the same clade," 2020.

[15] M. Shereen, S. Khan, A. Kazmi, and N. Bashir, "COVID-19 infection: Origin, transmission, and characteristics of human coronaviruses," 2020.

[16] B. Cohen and H. Hull, "Study Guide for Memmler's the Human Body in Health and Disease," 2020.

[17] M. Bizzoca, G. Campisi, and L. Muzio, "Covid-19 Pandemic: What Changes for Dentists and Oral Medicine Experts? A Narrative Review and Novel Approaches to Infection Containment," 2020.

[18] G. Ghartey-Kwansah, Q. Yin, and Z. Li, "Calcium-dependent Protein Kinases in Malaria Parasite Development and Infection," 2020.

[19] Y. Jin, L. Cai, Z. Cheng, H. Cheng, and T. Deng, "A rapid advice guideline for the diagnosis and treatment of 2019 novel coronavirus (2019-nCoV) infected pneumonia (standard version)," 2020.

[20] M. Wujtewicz and A. Dylczyk-Sommer, "COVID-19: What should anaethesiologists and intensivists know about it?" 2020.

[21] M. Bizzoca, G. Campisi, and L. Muzio, "Covid-19 Pandemic: What Changes for Dentists and Oral Medicine Experts? A Narrative Review and Novel Approaches to Infection Containment," 2020.

[22] J. Pang, M. Wang, and S. Tan, "Potential rapid diagnostics, vaccine and therapeutics for 2019 novel coronavirus (2019-nCoV): A systematic review," 2020.

[23] P. Diasso, H. Birke, and S. Nielsen, "The effects of long-term opioid treatment on the immune system in chronic non-cancer pain patients: A systematic review," 2020.

[24] M. Puig-Domingo, "COVID-19 and endocrine diseases. A statement from the European Society of Endocrinology," 2020.

[25] K. Narin and Z. Pamuk, "Automatic detection of coronavirus disease (covid-19) using x-ray images and deep convolutional neural networks," 2020.

[26] Y. Guo, Q. Cao, Z. Hong, Y. Tan, and S. Chen, "The origin, transmission and clinical therapies on coronavirus disease 2019 (COVID-19) outbreak: An update on the status," 2020.

[27] N. Khan, S. Fahad, and M. Naushad, "COVID-2019 Locked down Impact on Dairy Industry in the World," 2020.

[28] J. Nikolich-Zugich, S. Knox, C. Rios, and B. Natt, "SARS-CoV-2 and COVID-19 in older adults: What we may expect regarding pathogenesis, immune responses, and outcomes," 2020.

[29] H. Legido-Quigley and J. Mateos-García, "The resilience of the Spanish health system against the COVID-19 pandemic," 2020.

[30] L. Liu, Q. Zhou, Y. Li, V. Garner, S. Watkins, and G. Carter, "Research and development on therapeutic agents and vaccines for COVID-19 and related human coronavirus diseases," 2020.

[31] H. Li, Z. Yao, Z. Zhang, X. L. Cai, G. Liu, G. Liu, and L. Cui, "The progress on physicochemical properties and biocompatibility of tantalum-based metal bone implants," 2020.

[32] Z. Udwadia, "How to protect the protectors: 10 lessons to learn for doctors fighting the COVID-19 coronavirus," 2020.

[33] J. Chan, S. Yuan, K. Kok, KKW, H. Chu, and J. Yang, "A familial cluster of pneumonia associated with the 2019 novel coronavirus indicating person-to-person transmission: A study of a family cluster," 2020.

[34] Y. Wang, Y. Wang, and Q. Chen, "Unique epidemiological and clinical features of the emerging 2019 novel coronavirus pneumonia (COVID-19) implicate special control measures," 2020.

[35] H. Rothan, "The epidemiology and pathogenesis of coronavirus disease (COVID-19) outbreak," 2020.

[36] Z. Zhang, F. Yu, Y. Zou, A. Qiu, and T. Wu, "Phage protein receptors have multiple interaction partners and high expressions," 2020.

[37] D. Moreira, M. Pereira, A. Lopes, and S. Coimbra, "The best CRISPR/Cas9 versus RNA interference approaches for Arabinogalactan proteins study," 2020.

[38] L. Keeble, N. Moser, and J. Rodriguez-Manzano, "ISFET-based Sensing and Electric Field Actuation of DNA for On-Chip Detection: A Review," 2020.

[39] Y. Rimal, "Boruta Algorithm is significant for large feature selection of student marks data of Pokhara University, Nepal," 2020.

[40] L. Blairon, A. B. I. Wilmet, and N. Tré-Hardy, "Implementation of rapid SARS-CoV-2 antigenic testing in a laboratory without access to molecular methods: experiences of a general hospital," *Journal of Clinical Virology*, 2020.

[41] F. Capello, N. Cipolla, L. Cosco, and Gnasso.A, "The VivaDiag COVID-19 lgM/IgG Rapid Test for the Screening and Early Diagnosis of COVID-19 in patients with no clinical signs of the disease," 2020.

[42] Y. Rimal, "Deterministic Machine Learning Cluster Analysis of Research Data: Using R Programming," 2020.

6

COVID Cases Analysis: Dynamic Animated Plots Using R Programming

Yagyanath Rimal, Bharatendra Rai, Vijay Singh Rathore, and Sakuntala Pageni

CONTENTS

6.1 Introduction

Data visualization primarily refers to the graphical representation of data using charts, graphs, and maps to provide an understanding of trends and patterns of data [1]. However, there are more tools available in the modern world for analyzing multidimensional data. R provides data-master crunching and analysis with graphical animation with easy and quick plots. The developers of the r project were Robert Gentleman and Ross Ihaka in 1996 for static interpretation of data, easy data wrangling, advanced visualization, open sources advance, and development. The function ggplot2, gg-animation, plot and geom data map are the best tools for large data analysis like COVID cases these days. The outbreak of COVID-19 caused more than 11 million infections and 500,000 deaths of humans within a few months of pandemic's start in late 2019 [2]. The coronavirus infects living cells, which contain genetic materials that spike protein could easily replicate to more viruses and outer surfaces with protein spikes when patients cough or sneeze, spreading droplets through the air. When a virus in the heel, through the nose, mouth, and throat of lungs then spike of virus contracts with perceptual molecules of lungs cell membrane of the spike protein of the virus [3] [4]. [5] [6]. The virus gets into cell ribosome of cell components in the lungs and ultimately multiplies inside lungs with the blood of oxygen and transmutes them to capillaries inside a sac of lungs purified inside alveoli. When the human immune system is healthy, carbon dioxide and [7] virus directly inserted into lungs and multiplying create pneumonia, with water in lungs, which leads to difficulty breathing. This process causes pneumonia, headache, and increased difficulty breathing in many areas of the lungs [8] [9]. More serious patients need a ventilator to breathe. However, this may vary according to age, previous diseases, the medicine used, and many more factors [10]. The data visualization and summary of the ongoing pattern of COVID cases provide better guidelines for taking preventive measures for the local, national and international planner.

The spread of coronavirus begins at Wuhan's live animal food market, then to every country, which starts the economies going down, and hospital go down suddenly with their failing hospital systems facing every country in the world [11]. By March 5, 2020,, there were 95,000 cases worldwide, with 3,300 deaths in 65 countries, and within a single month it increased to 2,921,439 cases with a highly accelerating death rate rising to 203,300 around the world. Many were losing their lives every day and the world was crying out for health. The United States had the highest number of deaths, 24,000, and there were more than 600,000 cases, with an increasing number of 1,500 each day, causing crisis situations. Similarly, Spain reached 159,516 cases, with 20,465 deaths [12]. According to the WHO report the fatality rate (CFR) of coronavirus has mainly varied, following the previous history of other diseases conformed to death [12]. Similarly, the age of the sick varies from young to old with an immunity rate that decreases with age.

The above two-column describes the age-wise fertility rate, which implies the older the age the more the risk. The elderly person's risk increases if a person of the second two-column describes that the previous disease suf-. ferer has a higher chance of being diseased [13]. So, morally, the pertinent was not high risk and there was no report that corona virus-infected children by transmission or infant nursing [10] [14]. While breastfeeding, the main risk to mother and children is in droplets rather than in the breastfeeding. At present, 80 percent of patients had mild infections, and some people did not develop symptoms at all [6] [15].

There were many tests behind the pretest of COVID cases, including the walking test [9] [16]. If the patient feels a reduction of oxygen, then one should go get a doctors' advice; this process is known as happy hypoxemia. When decreasing the oxygen level and the lungs will not take more oxygen from the atmosphere. Then the patient will be on anti-inflammatory drugs like HCQ, antiviral medicine like remdicevir, and heavy flies and monoclonal medicines are being prescribed [17] [18]. The blood thinner medicine and storied were added based on a patient blood tests with a hyper cognitive state. Generally, those patients who have more than 60 years of kidney, heart diseases were recommended to be hospitalized immediately [19]. There were mild, moderate, and severe cases, too. If the patient with more symptoms for not long three days fell in the mild category, whereas a patient with pneumonia with fever and chest pain was termed as moderate, while the elderly with diseases and fever were considered severe cases. The mildly infected patient was recommended to stay home in isolation, monitoring fever and oxygen levels. Moderate and severe patients having more respiratory problems were recommended to admit. It is generally recommended that close contact of less than 3 feet and for more than 15 minutes means meetings with COVID patients will transmit the virus [20]. The RTPCR test done with the nasal swab is the confirmatory test when first meeting with COVID close contact within 5 to 10 days after. These test results have more specificity and sensitivity of prediction [15]. At the time of discharge after 14 days' treatment and the patient having no fever for the previous three days, it is not strongly recommended to test again when checking out from the hospital. If tested it will result in one more positive due to the deadly virus on the swab. Although there is an eight-week test positive found after recovering, the body already had great immunity to fight against COVID-19 diseases [21]. Therefore, COVID is a disease that is the more problematic for those who have had diabetes, kidney, and cardiovascular diseases already. The social distancing with mask-wearing, hand washing, and avoiding crowded places were very common preventive measures [22]. However, some countries unlock increases the more COVID cases found each day. Thus, in analysis of COVID cases by animated charts, plots are of significant importance for quickly recommending further action implementation in the current situation. The static data only presents facts, but not trends and patterns of COVID cases by region [23].

Therefore, this exploratory analysis tries to develop animated trends of COVID cases for individual as well as global trends.

6.2 Method

The owid- COVID -data combines ECDC data on confirmed cases and deaths with the testing data collected daily and verified sources such as Johns Hopkins University. The data repository includes ios code, continent, location, total cases, total population, and much more in 40 columns of information related to the worldwide pandemic context. The 40,959 observations were describing many properties of COVID data as.

However, data presented in tabular format is not able to describe overall patterns and trends of world data without some type of visual comparison

TABLE 6.1

Format of Data Source after Download

Sn	iso_code	continent	location	date	total_cases	new_cases
1	ABW	North America	Aruba	3/13/2020	2	2
2	ABW	North America	Aruba	3/19/2020		
....						
40749			International	31/08/2020	696	7

at first glance of data. Therefore, the present chapter tries to explore COVID case comparison and analysis with many dimensions of data trends of the day and monthly patterns of world COVID. Data management requires machine learning [24], for the animation table of data is converted as a data frame with converting its data with the character format of the day, month, and year pattern in the date column. The mutate command is used to change its data name by sub-situating another pattern under the lubricate library. Whose day, month, and year data could easily convert from the original date to another user-defined date (Y-M-D). The final date, which becomes the 2019-12-31 pattern, could separate its pattern too. The animated line chart displays segments of series between two points with time intervals such as days, months, and years for better analysis of the COVID daily record of trending patterns. After converting data format datanew=data %>% mutate(date=dmy(date)) and new2 %>% group_by(date) %>% summaries (count=n ()) %>% mutate (Cuml=cumsum(count))

6.3   Results

The data, with 40,748 observations, could be grouped by date and calculated. Its cumulative sum of each new date summarizes its output in Tibble and generates observations of 199 countries with 3 variables: date, count, and Culm, as below.

$ date: Date [1:245], format: "2019-12-31" "2020-01-01" ...

$ count: int [1:245] 68 68 68 68 68 68 68 68 68 68 ...

$ Culm: int [1:245] 68 136 204 272 340 408 476 544 612 680 ...

Date count Culm

	Date	count	Culm
1	2019-12-31	68	68
2	2020-01-01	68	136
....			
245	2020-08-31	211	40749
246	2020-09-01	210	40959

6.3.1   Line Plot Animation

After converting data format into table format the animated plot generates using ggplot with line geom points with filling red color with transition by cumulative sum by the day's records. yy %>% ggplot (aes (x=date, y=Cuml)) + geom_line(color='red') + geom_point(size=1.5) geom_area(fill= 'red') + theme_bw () + ggtitle ('Daily Cumulative Cases') transition reveal (Culm). The animated line graphs of COVID indicates its day ration have been increasing daily.

The daily cases of worldwide COVID were gradually increasing by months whose linearity adverse sharpen after last week from March. The color and its line animation could be defined as require reached more than 257,000 new cases per day. The monthly line plot of a selected country could be easily filtered using month is equal to 2 and, group by day, country name, and their total cases of each month stored in data of COVID records. df=data new%>% filter(m==2) %>% group by (d, location, total cases) %>% summaries (count= n ()) %>% mutate (Culm=cum-sum (total cases))

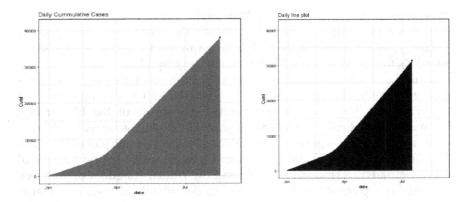

FIGURE 6.1
Daily Cumulative Cases Animation Plot.

This command summarized 116 records of four countries, whose count and
the cumulative new cases could calculate and store in the same row whose
complete cases were stored otherwise counted values replaced by 0. This
new=data. frame (complete (df, d, location, fill =list(count=0))). Similarly,
using the same data sets with the selected country could filtered using or
command like new2=new %>% filter (location == 'United States' | location =
= 'India' | location == 'United Kingdom' | location == 'Germany' | location
== 'Brazil' | location == 'Nepal') for the purpose of daily line graphs of the
selected country. This record displays 200 cases of the selected country with
cumulative cases.

d	location	total cases	count	Culm
1	France	6	1	6
1	Germany	7	1	7
1	United Kingdom	2	1	2
1	United States	7	1	7
2	France	6	1	6
2	Germany	8	1	8
2	United Kingdom	2	1	2
....				
25	India	3167323	1	3167323
25	Brazil	3622861	1	3622861
25	Nepal	32678	1	32678
25	Russia	961493	1	961493
25	United States	5740909	1	5740909

These types of records could plot inline graphs using ggplot (data=new2, aes
(x=d, y=Culm, group=location, color=location)) + germline () + geoponic ()
+ scale_y_log10() + theme_bw () + tittle ('Animated line plot') + transition
reveal(d). this command quickly generates line graphs.

Similarly, when data filter on June with more countries with the comparison animated line graph Similarly, when data filter on June with more
countries with the comparison animated line graph of the US, France,
Germany, India, Russia, Brazil, and Nepal country summarize with 245 items
with cumulative cases generates the table as below. This line plot of February
records shows that the United States and India have the lowest number of
cases compared to France, Germany, and the United Kingdom of both two
animated line plots when the second plots describe the day-wise animated
plots of July and August 2020 explores India has sharply increased and Nepal
has still gradually in comparison with the top country. This model provides
the facility to generate the animated plots of a single month as the user
required with legend at right. From the above line plots, the COVID cases of
France, Germany, and the UK present with more stable conditions, whereas

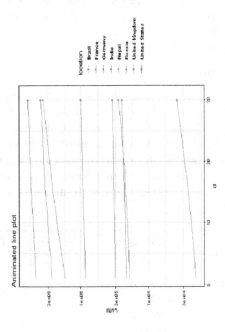

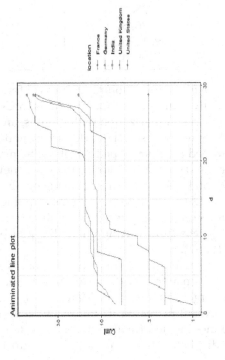

FIGURE 6.2

Animinitate Line Plot of Developed Country.

Brazil, the United States, and India have the largest new case records per day, implying that the government and citizens of these countries should take more precautions for COVID control measures.

6.3.2   Bar Plot Animation

The bar plot graph is used to present categorical in rectangular bars proportional to COVID cases vertically or horizontally with day, month, and year. The time pattern for better analysis could be designed after filtering selected country records of the fifth or sixth months' records summarized as follows: new=data new %>% filter (location =='United States' |location == 'France' |location == 'United Kingdom'| location == 'Germany') %>% filter(m==5|m==6) %>% group by (location, m, total cases) %>% summaries (count=n ())

d	location	total cases	count	Culm
1	France	151753	1	151753
1	United Kingdom	274762	1	274762
1	United States	1790191	1	1790191
1	Germany	182028	1	182028
2	France	152091	1	152091
25	India	3167323	1	3167323
25	Brazil	3622861	1	3622861
25	Nepal	32678	1	32678
25	Russia	961493	1	961493
25	United States	5740909	1	5740909

This output filtered the May and June monthly records of four selected countries only, with a cumulative sum of new COVID cases in Culm columns of 245 observations. After data summarization with months, any months starts COVID can be selected. Here, the researcher selects May and June records of world COVID data grouped by a developed country. The map records with a bar diagram with log values filled by country name. p= new %>% gwplot (aes (x = location, y = total cases, fill = location)) geom_bar (stat = 'identity') geom_point(size = 1.5) scale_y_log10() theme () guides(fill = F), and this data will be transits with monthly records of a time frame as p +transition time (as. integer(m)) labs (title = 'Animinitate Bar plot by Month,' subtitle = 'Month: {frame_time}').

When the filter the data using or operator of all months since starts the animated plots describes its pattern of increment of each country. The United States and India have largely increased in March 2020, and they became the top countries until Nepal increased its ratio in June and July. This bar chart presents the two months' information of four selected country record bars whose animation display rises according to daily information for each

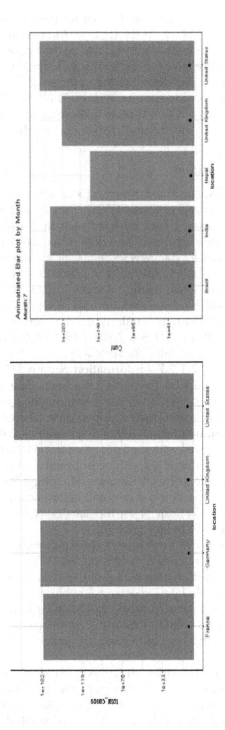

FIGURE 6.3
Animated Bar Plot.

country. Similarly, the total cases could be easily plotted in bar graphs as p+ transition states (total cases) + labs (title = 'Animation bar plot of top countries') shadow mark () + enter grow () and p+ transition states (Culm)+ labs (title = 'Animation bar plot for top countries') shadow mark () + enter grow (). The animation bar plot describes their pattern of COVID cases of the fifth and sixth months' records simultaneously. The three European countries have a similar pattern of COVID cases in these months, with the United States increasing sharply at the end and crossing over in June 2020.

6.3.3   Bubble Plot Animation

The bubble chart that displays three dimensions of data on x and y axis with plotted data on time intervals of total COVID cases data of month and day records is plotted. After loading data, the head(data) and str(data) commands describe the data and their structure. The data set converts its data pattern using lubricate packages datanew$date=as. Date(datanew$date/%d/ %Y") convert its date as [1] "12/31/2019" "1/1/2020" "1/2/2020" "1/3/ 2020" "1/4/2020" pattern when using data new = data %>% mutate(date= mdy(date)) the data will be converted into [1] "2019-12-31" "2020-01-01" "2020-01-02" "2020-01-03" pattern. Similarly, the day records datanew$date= day(datanew$date) [1] 31 1 2 3 4 5 6 7 [985] 15 16 4 6 7 8 and month > datanew$m=month(datanew$date) converts as [1] 12 1 1 1 1 1 1 6 6 6 6 6 6 6 6 6 7 7 7 7 7 7 7 7 7 7 7 7 7 7 7 7 information pattern. Whose data again data2=datanew %>% group by (location, continent, date, GDP_ precipitate, life expectancy, total cases, m, d) %>% summaries (count=n ()) %>% mutate(total=consume(count)) formed grouped data. The group by records with country name, continent, day, GDP, life acceptancy with months and day records were summarized with total new records on total columns stored in new table. Thus, the bubble plots with consideration with life expectancy and per capita of population of COVID cases. The p=ggplot (data2, aes (x= gdp_per_capita, y=life expectancy, size=total, color=location)) geom_point (show. legend=F, alpha=0.7) scale_x_log10() labs (x='GDP Per Capita', y= 'Life Expectancy') scale size (range=c (2,15)).

The base bubble plots GDP versus life expectancy with total cases plots the bubble very small to large when increasing the COVID cases of all countries. When applying transition time with month p+transition_time(m)+ labs (title = 'Year per Capital Vs Life Expectancy', subtitle='year: {frame time}') shadow wake (0.5). this output could be easily saved into specified direction anim_save ('C:/Users/Rimal/Desktop/cove map/month plot'). With saving animated plot so that we could easily be embedded into PowerPoint slides.

The first map with the transition described was made on January 2020 information and demonstrates there were very few countries recording the few COVID cases, whereas in July of the second plot almost all countries in the world were being infected by COVID-19; the small dots indicate the

FIGURE 6.4
Animated Bubble Plot.

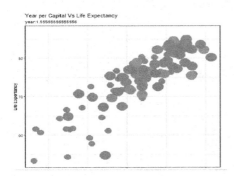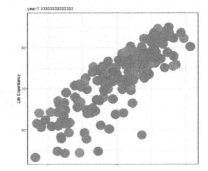

FIGURE 6.5
Animated Bobble Plot of COVID Cases.

few cases proportion with a population of 3D animated bubble plots. The bubble plot when time transition with date animates all continents cases of 3D notion. Using p+ transition _time(date)+ labs (title='GDP VS Life', subtitle = 'Year: {frame time}') shadow wake (0.5) + facet wrap (~continent) anim_save ('C:/Users/ /Desktop/ covid map/ continent').

The first continental maps represent that the data of Africa and Asia show the highest number of cases, while South America and Oceana have three countries with very few cases. At the end of Jul,y almost all continents were being infected with more COVID cases. Thus, the above line, bar, and bubble animated plots give more information of COVID cases whose daily and monthly records were steadily increasing.

6.4 Conclusion

By 2021, COVID-19 had become the most dangerous threat to human civilization. Holistic approaches are required for COVID disease control, and

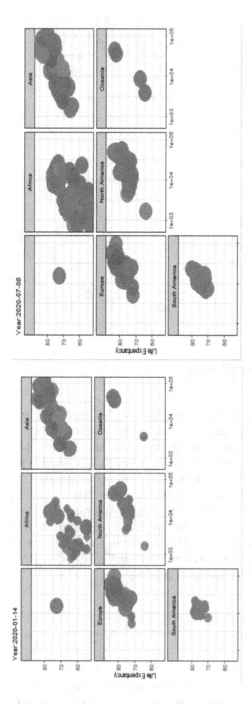

FIGURE 6.6
Animated Continent Bubble Plot.

WHO collects datasets about coronavirus from regions, countries, cities, and many patterns of data visualization. However, there is a huge potential to analyze data with dynamic patterns and trends study for a further new strategy to control measures. The maps describe the relationship of provinces within one country and neighboring provinces in another country. For better management planning, dynamic plots like line, bar, and bubble describe the more ongoing COVID cases to determine the relationship between province cases for better preventive measures. The United States, Brazil, India, and Russia were top countries where new cases increased daily. Therefore, further planning with dynamic animated data can play a significant role in the battle against COVID-19's spread around the world.

References

[1] C. R. MacIntyre, Public health, health systems and palliation planning for COVID-19 on an exponential timeline, 2020.

[2] S. L. BE Bannish, Exploring Modeling by Programming: Insights from Numerical Experimentation, 2020.

[3] J. Hick, D. Hanfling, M. Wynia, and A. Pavia, Duty to plan: Health care, crisis standards of care, and novel coronavirus SARS-CoV-2, 2020.

[4] R. Riet and K. v. Dijk, New possibilities on transplanting kidneys from hepatitis C virus positive donors: A systematic review, 2020.

[5] G. Onder, G. Rezza, and S. BrusaferroJama, Case-fatality rate and characteristics of patients dying in relation to COVID-19 in Italy, 2020.

[6] K. S., COVID-19: A brief clinical overview, *Journal of Geriatric Care and Research*, 2020, 2020.

[7] R. Adury, V. Chowdhuryl, A. Vijay, and R. Singh, Generate, repurpose, validate: A receptor-mediated atom-by-atom drug generation for SARS-Cov-2 spike protein and similarity-mapped drug repurposing, 2020.

[8] Z. Ye, S. Yuan, K. Yuen, and S. Fung, Zoonotic origins of human coronaviruses, 2020.

[9] B. Chhikara, B. Rathi, and J. Singh, Corona virus SARS-CoV-2 disease COVID-19: Infection, prevention and clinical advances of the prospective chemical drug therapeutics, 2020.

[10] Q. He, H. Liu, C. Huang, R. Wang, and M. Luo, Herpes Simplex Virus 1-Induced Blood-Brain Barrier Damage Involves Apoptosis Associated with GM130-Mediated Golgi Stress, 2020.

[11] T. A. El-Aziz and J. Stockand, Recent progress and challenges in drug development against COVID-19 coronavirus (SARS-CoV-2) – an update on the status, 2020.

[12] W. McKibbin and R. Fernando, The global macroeconomic impacts of COVID-19: Seven scenarios, 2020.

[13] M. Battegay, R. Kuehl, S. Tschudin-Sutter, and H. Hirsch, 2019-novel Coronavirus (2019-nCoV): Estimating the case fatality rate–a word of caution, 2020.

[14] P. Lunn, S. Timmons, M. Barjaková, and C. Belton, Motivating social distancing during the Covid-19 pandemic: An online experiment, 2020.

[15] P. Lunn, S. Timmons, C. Belton, and M. Barjahova, Motivating social distancing during the Covid-19 pandemic: An online experiment. ESRI Working Paper No. 658 April 2020, 2020.

[16] A. L, L. B, E. Guyen, M. L, M. D, F. R. Herrmann and S. A., Effect of different walking aids on walking capacity of patients with poststroke hemiparesis. *Archives of Physical Medicine and Rehabilitation*, 90(8), 1408–, 2009.

[17] Y. Wu, C., Liu, C. Yeh, L. Sung, C. Lin, and Y. Cherng, Hospitalization outcomes of heart diseases between patients who received medical care by cardiologists and non-cardiologist physicians: A propensity score, 2020.

[18] L. Poon, H. Yang, A. Kapur, and N. Melamed, Global interim guidance on coronavirus disease 2019 (COVID-19) during pregnancy and puerperium from FIGO and allied partners: Information for healthcare …, 2020.

[19] A. M and W. G., Fast curvelet transform through genetic algorithm for multimodal medical image fusion, *Soft Computing*, pp. 1815–1836.

[20] L. Friedman, COVID-19 Testing Data from Our World in Data, https://data.world/liz-friedman/covid-19-testing-data-from-our-world-in-data, 2020.

[21] Z. Kelly, Powerful Benefits of Using R to Analyze Your Research Data (and a Few Limitations), 2018.

[22] T. Doron, COVID-19 Rapid Response Research with Workflow-based Data Analysis, 2020.

[23] R. Yagyanath, Machine Learning Techniques Best for Large Data Prediction A Case Study of Breast Cancer Categorical Data: k-Nearest Neighbors, *Ontology-Based Information Retrieval for Healthcare Systems*, p. 245, 2020.

[24] R. Yagyanath, G. Saikat, and B. Aakriti, Data Interpretation and Visualization of COVID-19 Cases using R Programming, *Informatics in Medicine*, 2021.

7

Tracking and Analyzing COVID-19 Pandemic Using Twitter and Topic Modelling

Pradeep Gangwar, Yashbir Singh, and Vrijendra Singh

CONTENTS

7.1 Introduction

COVID-19 left the whole world trapped inside their houses. With a lot of confusion among the general public, it was time for researchers to study different aspects of the disease in order to find an effective way out of it. With high infection rates it was difficult to trace the virus as its transmission was way ahead of what everyone thought. Penetrating into new cities every day, it trapped the whole world very quickly. Experts suggested that contact tracing, isolation and tracking the virus were among the most effective solutions to reduce its spread.

Can we use artificial intelligence to trace and track COVID-19 spread? Well, the answer is yes. With new research emerging every day in this field, artificial

DOI: 10.1201/9781003126218-7

intelligence has brought the world to a whole new level. From self-driving cars to smartphone assistants, artificial intelligence has made humans more dependent on machines than ever. In this chapter we will learn how we can use topic modelling, one of the famous machine-learning techniques to track and trace COVID-19 on Twitter.

7.2   What is Topic Modelling?

Topic modelling is an unsupervised machine learning technique that is capable of detecting patterns in words and phrases from a set of documents and clustering word groups and similar expressions that best characterize a set of documents. In big corporations, data analysis of such a huge set of documents would be a cumbersome task if delegated to humans. This is why everyone is excited about the implications of artificial intelligence in their day-to-day tasks and business as a whole. AI-powered text analysis uses a wide range of methods and algorithms to process language naturally, one of which is *Topic Analysis* – detecting topics from text.

By using topic analysis methods, businesses can offload such heavy tasks to machines instead of overloading employees with huge chunks of data. These techniques are helpful in organizations that receive huge amounts of queries/complaints on a daily basis. Routing such requests/queries/complaints from customers to respective departments is crucial to resolving them immediately.

7.3   Related Work

Topic analysis techniques are very popular among researchers, and these techniques have proven to be beneficial in various epidemics and disasters. As expected, researchers have done a good amount of work in this domain and have used various Natural Language Processing (NLP) techniques to study people's behavior during this pandemic. Here are a few research papers in this domain:

1. Dynamic Topic Modelling of the COVID-19 Twitter Narrative among US Governors and Cabinet Executives [1].
2. Tracking Public and Private Responses to the COVID-19 Epidemic: Evidence from State and Local Government Actions [2].

3. Exploratory Analysis of Covid-19 Tweets Using Topic Modelling, UMAP, and DiGraphs [3].

4. An Exploratory Study of COVID-19 Information on Twitter in the Greater Region. [4].

7.4 Other Ways of Topic Analysis

The other powerful technique in topic analysis is *Topic Classification*. While Topic Modelling is unsupervised technique, Topic Classification is supervised topic analysis technique. Models are trained on a set of data and provide better accuracy than Topic Modelling when tested on real data. Most of the mature businesses dealing with a specific set of data prefer supervised models because of their better accuracy.

While both of the methods, that is, Topic Modelling and Topic Classification, are useful, they have their own advantages and disadvantages. For example, if we are not sure if we will encounter some new information in future with a completely new topic, then Topic Modelling is a better choice because it can cluster new information together, but Topic Classification would provide less accuracy because it was not trained on those new topic models and never prepared for them to occur.

7.5 How Can Twitter and Topic Modelling Be Used in Tackling COVID-19?

7.5.1 Role of Twitter

With more than 300 million active monthly users, Twitter is one of the most popular microblogging sites across the world. On the other hand, Twitter is a microblogging service where users can broadcast short 140-character messages called *tweets*, which can be personal thoughts or opinions on public statements, places, persons, disasters, and other things. making it a useful tool for gathering information. Twitter has a presence in mostly all government departments, world leaders' offices, organizations, actors, and so forth, who actively share news/articles/announcements with their followers. During COVID-19 we saw a huge inflow of tweets regarding the COVID-19 spread, where government bodies were actively seen spreading awareness, sending advisories, and daily statistics to reach the public.

With so much data available at our disposal, researchers believe that this could be very beneficial if used properly. One study by Aramaki and colleagues [5] made use of the tweets of Japanese Twitter users to detect if an influenza epidemic is happening and predict what type of influenza will spread in any given season. Another study described an approach that automatically identified tweets that contribute to situational awareness [6].

7.5.2   Twitter and Covid

Social media, especially Twitter, has recently played an important role in natural disasters as an instrument of information. Such huge data if used properly can be used in managing and tracing the COVID-19 virus. Most governments have set up dedicated telephone helplines for their citizens and have deployed special portals and mobile applications for contact tracing. During the coronavirus most people sought help using twitter as one of the mediums. In initial days most common queries involved the available treatment, lockdown advisories, queries regarding COVID symptoms, queries regarding international travel, and so forth.

With such huge volumes of incoming requests daily it could have been overwhelming for any authority to respond to each of the queries. This is where topic modelling can be beneficial.

7.5.3   How Can Topic Modelling Help?

As we know, topic modelling is a method to analyse a set of documents (here, Tweets), cluster them into different groups based on their similarity, and then assign them some relevant topics.

Using such techniques, we can use incoming Tweets as an information stream and, after clustering them into different groups, we can forward them to respective authorities for quick rectification. This would help authorities identify similar requests from thousands of incoming tweets and quickly address them.

Researchers have been working on such techniques and have done impressive work in this area. We found some interesting work in this domain given below:

1. To track dengue epidemic using twitter content classification and topic modelling [7].

2. Using Topic Modelling to Make Sense of Typhoon-related Tweets [8].

3. Towards detecting influenza epidemics by analyzing Twitter messages [9].

7.6   How Does Topic Modelling Work?

The process involves different steps. Some of the steps are the same as other NLP algorithms. Given below are some of the steps involved in this process:

1. Data collection
2. Data cleaning
3. Topic modelling
4. Summarization

7.6.1   Data Collection

Data is collected from publicly available tweets using Twitter Public API. Tweets are fetched using some specific keywords, hashtags, and accounts that are relevant for the information we are looking for.

7.6.2   Data Cleaning

Data cleaning is an important step where data is cleaned to avoid producing misleading results. Tweets are checked for unwanted symbols, noise, and irrelevant data. Such unwanted data is then removed and everything is converted to lowercase.

7.6.3   Topic Modelling

Topic modelling refers to dividing corpus of documents into:

- A set of topics covered by documents in corpus.
- A different set of documents from the corpus grouped together based on the topic they cover.

The many methods that are used for topic modelling:

1. Latent Semantic Analysis (LSA)
2. Latent Dirichlet Allocation (LDA)
3. Non-Negative Matrix Factorization

Apart from these basic methods, researchers use other methods that are variants of these methods.

7.7   Further Study

An article on Monkey Learn about "Introduction to topic modelling" [10] has very detailed information about topic modelling, its variations, its applications and how it works. It also explains LSA and LDA in detail. It is a very good read for interested readers.

Others who wish to explore topic models in detail, helpful editorial information is available at Elsevier with title "Introduction—Topic models: What they are and why they matter" [11]. It should be a good read.

References

[1] Hao Sha, Mohammad Al Hasan, George Mohler, P. Jeffrey Brantingham, Dynamic topic modeling of the COVID-19 Twitter narrative among U.S. governors and cabinet executives. arXiv:2004.11692 (2020).

[2] Sumedha Gupta, Thuy D. Nguyen, Felipe Lozano Rojas, Shyam Raman, Byungkyu Lee, Ana Bento, Kosali I. Simon, and Coady Wing. Tracking Public and Private Responses to the COVID-19 Epidemic: Evidence from State and Local Government Actions, NBER Working Papers 27027, National Bureau of Economic Research (2020).

[3] Catherine Ordun, Sanjay Purushotham, and Edward Raff. Exploratory Analysis of Covid-19 Tweets using Topic Modeling, UMAP, and DiGraphs. arXiv:2005.03082 (2020).

[4] Ninghan Chen, Zhiqiang Zhong, and Jun Pang. An Exploratory Study of COVID-19 Information on Twitter in the Greater Region. arXiv:2008.05900. August 2020.

[5] E. Aramaki, S. Maskawa, and M. Morita. Twitter Catches the Flu: Detecting Influenza Epidemics Using Twitter. In *Proceedings of the 2011 Conference on Empirical Methods in Natural Language Processing*. July 27–31, 2011.

[6] S. Verma et al., Natural Language Processing to the Rescue? Extracting "Situational Awareness" Tweets During Mass Emergency. In *Proceedings of the Fifth International AAAI Conference on Weblogs and Social Media*. 2011.

[7] Missier P. et al. (2016) Tracking Dengue Epidemics Using Twitter Content Classification and Topic Modelling. In: Casteleyn S., Dolog P., and Pautasso C. (eds) *Current Trends in Web Engineering. ICWE 2016.* Lecture Notes in Computer Science, vol 9881. Springer, Cham. https://doi.org/10.1007/978-3-319-46963-8_7.

[8] C. Ligutom, J. V. Orio, D. A. Marie Ramacho, C. Montenegro, R. E. Roxas, and N. Oco, Using Topic Modelling to make sense of typhoon-related tweets, 2016 International Conference on Asian Language Processing (IALP), Tainan, 2016, pp. 362–365, doi:10.1109/IALP.2016.7876006.

[9] Aron Culotta. 2010. Towards detecting influenza epidemics by analyzing Twitter messages. In *Proceedings of the First Workshop on Social Media Analytics*

(*SOMA '10*). Association for Computing Machinery: New York, 115–122. DOI: https://doi.org/10.1145/1964858.1964874.

[10] Topic Modelling Introduction https://monkeylearn.com/blog/introduction-to-topic-modeling/

[11] John W. Mohr, PetkoBogdanov. Introduction – Topic models: What they are and why they matter. https://doi.org/10.1016/j.poetic.2013.10.001 (Dec., 2013)

8

Artificial Neural Network Application to Analyze 3D Image Printing Using Artificial Intelligence in COVID-19

Paryati and Salahddine Krit

CONTENTS

DOI: 10.1201/9781003126218-8

8.1   Introduction

8.1.1   Background

In general, the important aspect underlying several theories in artificial intelligence is the design-recognition system. The design-recognition concept can be implemented in several sector, including the medical sector, military sector, education, and so forth. The principal problem in design-recognition implementation is how the data acquisition is executed so that the amount of numerical data produced is representative and consistent about the given sample. In this research, the writer makes counter propagation neural network application for acquaintance in the three-dimensional image shape's design, with a result that can be identified by recognizing the design method. The media for save data using Microsoft Visual Basic version 7.0 and Microsoft Access.

8.1.2   Problem Formula

Building on the introduction, a problem is formulated by the method of how to make design recognizable for three-dimension image application, which is an implementation from the design recognize concept.

8.1.3   Problem Limit

This problem will limited by:

1. Image, which can be processed, is a drawing image with tool prepared from the application.
2. Criticize the extraction data method from a digital image becomes series of numeric data.

8.1.4   Direction

Make counter propagation neural network application for acquainted in three dimension image shape's design with the result that can be identified by design recognize design method.

8.1.5   Divining Annual Research

The extraction data method for recognizing character and the numbered hand-written note using counter propagation neural network application, are related in a certain way. The difference with this research is the conversion image aimed to be the input of the counter propagation neural network and extraction data method, especially for three-dimensional image shape.

8.2 Literature Review

8.2.1 Image Processing Substance

Generally, image is defined as a visual representation of an object. In the computer field, the image is a visual representation from an object after experiencing several data transformations from several series of numeric.

8.2.2 Counter-Propagation Neural Network

Counter-propagation is one of the artificial neural network's learning process where, in the process changing, value is backward – that is from the output layer and finally the input layer. The algorithm is as follows:

1. Neuron output value calculation at hidden layer and output.

$$net_i = \sum_{i-1}^{n} (w_{ij}s_j) + q_i$$

$$s_i = f(net_i)$$

Where,

i: neuron number that is being calculated its activation signal
j: neuron number that the output is contributed to i neuron
s_j: output value j neuron
w_{ij}: relation content value between to i and j
q_i: neuron bias value to I net function or f (net$_i$) is also called "activation function," that the form can vary.

2. Mistake calculation in learning process is called energy function.

$$E = \sum_{(x,y)} \sum_{i} (y_i^x - s_i^s)^2$$

Where,

y_i : si I output neuron target output
x: actual output neuron output I when the network connected to x sample

3. Neuron (δi) sensitivity calculation in the hidden layers and output where the equation used to sensitivity calculation, both for hidden layers or output layers, depends on the activation function used.
4. Weigh and can change value calculation.

Weight change:

$$\Delta W_{ij}(x,y)\delta_i s_j^x$$

Can change:

$$\Delta q_i(x,y)\delta_i$$

5. New weight and can calculation.

New weight:

$$w_{ij}^{i+1} = w_{ij}^i + \Delta w_{ij} + momentum\ \Delta w_{ij}^{i-1}$$

New can:

$$q_i^{i+1} = q_i^i + \Delta q_i^1$$

6. The steps are repeated until a small output deviation reaches expected stopping criteria error.

This writing used is an important parameter of sigmoid activation functions other from JSB is how output from JSB is represented as distributive.

8.3   Discussion and Implementation

8.3.1   Case Analysis

An image that will be identified, called sample, must go through certain steps so there can be a good input. The inputs that can be well accepted are numeric data. Therefore, the case of how to converse a digital image becomes series of numeric data that are representative and consistent.

8.3.2   Data Acquisition Method

Each data sample that will be researched and analyzed must be well represented numeric data. Therefore, it needs a method that can consistently extract characteristic data from each sample.

8.3.3   Data Extraction

To obtain accurate and consistent data from the sample, a simple method used is counting active pixel numbers that are available in sample parts. The numeric data extraction algorithm from each sample is:

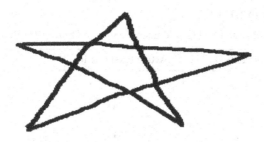

FIGURE 8.1
Sample that represents "star" image.

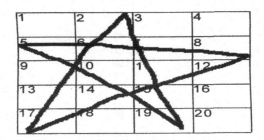

FIGURE 8.2
Region division of "star" sample.

1. Each researched sample is divided into several areas, such as four columns and five rows, so that will become 20 research regions.
2. Active pixel numbers (which is not white, but black) in each region is calculated accurately.
3. Obtained some 20 pieces of numeric data with column and row attributions that is expected to represent characteristics from expected samples.

Figure 8.2 shows region divisions on "star" image. Then, after the picture is divided into regions, the next steps calculate active pixels from each region. The result from active pixels calculation each region are expressed in Table 8.1.

8.3.4 Data Normalization

To maintain data consistency in each research sample, each numeric data must be through a normalization process. The normalization method that will be used is itself a very simple method that is fixed compared normalization. In this case is a comparison between active pixel numbers in each region with the most active pixels. The result from the normalization process is a series

TABLE 8.1

Active Pixel Grade Each Region in "Star" Sample

Region	Active Pixel Grade
1	1
2	82
3	3
4	1
5	74
6	111
7	81
8	87
9	76
10	71
11	91
12	90
13	66
14	101
15	126
16	1
17	112
18	23
19	89
20	1

of fractions rated from 0 (zero) to 1 (one). Table 8.1 shows that the highest pixel numbers lie in region 20, that is 118. Therefore, active pixel numbers from each region will be divided into 118 so that will result in numeric data as shown in Table 8.2.

8.3.5   Numeric Data Accumulation

Each sample of numeric data is normalized so that data collection will be accumulated in the spray form so that it is implemented in JSB. The accumulated final data must cover all input parameters needed by JSB from each parameter sample. The data that must be output is a target from each sample and also numeric data collection from each area is in each sample.

8.3.6   Artificial Neural Network Structure

JSB is used as a tool to analyze and also to proof the extraction concept that is made.

JSB Structure is a collection of in-order neurons that forms a meaningful structure. JSB structure as be seen. The following is neuron quantity rule in each layer can be seen in Table 8.3, below.

JSB has some algorithm that can be implemented. Here, a counter propagation algorithm is used. The activation function used is sigmoid activation

TABLE 8.2

Normalized Numeric Data Each Region in "Star" Sample

Region	Active Pixel Grade	
	Original	Normalized
1	0	1
2	92	0.7613
3	1	0.0082
4	0	1
5	84	0.6871
6	118	0.8918
7	81	0.6414
8	67	0.5439
9	60	0.4467
10	63	0.4752
11	84	0.6739
12	72	0.5788
13	53	0.3969
14	96	0.799
15	111	0.9366
16	0	1
17	119	1
18	27	0.174
19	91	0.7491
20	0	1

TABLE 8.3

Neuron Distribution

Layer	Neuron quantity
Input	depends on region quantity
Hidden	20
Output	18

other than structure, algorithm, activation function, initial value parameter, learn rate, momentum, and stopping criteria error, hard trapping, and match level will affect the final result of JSB.

8.4 Analysis and Simulation

8.4.1 Sample Preparation

The sample making needs to pay attention carefully to obtain consistent data towards target value. The form of the samples must be consistent and uniform. The sample quantity will also affect JSB performance.

8.4.2   Learning Activity Simulation

The first experiment done is a value-effect analysis of Learn Rate and Momentum and, finally, is initial value reach effect. For the three first experiments used initial value reach of 02 to 2. The experiment that will be done will be given several limitations so that can be obtained an effective result. The limitations are as follows:

1. Sample used is sample 1 with Region = 5*4;
2. Maximum epoch quantity is 1000;
3. Learning process is stopped if it has gained one of three criteria, that is, Hard Trapping, Match Level, or Stopping Criteria Error;
4. For each case several experiments are done and the best one is taken to be the temporary manual for the next learning process;
5. The best result of experiment from the biggest grade of Hard Trapping, Match Level and/or the smallest Stopping Criteria Error value;
6. The objective is to obtain the best JSB characteristic so that the JSB can "learn" pattern given the best.

8.4.3   Learn Rate Effect

To do Learn Rate Effect to JSB, first it must be decided that Momentum = 0, in order that the experiment will not be affected by the Momentum value. The experiment is done in critical and very extreme of 0.1, 0.3, 0.5, and 0.9.

From the five of the simulation, it can be concluded that the best Learn Rates by the lowest Error Level is 0.1, and the best Learn Rates based on the highest Match Level is 0.3. Otherwise, due to successful rate from this experiment is not measured by low rate of Error Level but from the high of Match Level rate, so that Simulation 2 with Learn Rates = 0.3 that will be used in the next experiment.

TABLE 8.4

Learn Rates Effect Experiment Result

Experiment	Learn Rate	Epoch	Error Level	Match Level
Simulation 1	0.1	1000	21.32567	45%
Simulation 2	0.3	1000	35.31232	58%
Simulation 3	0.5	1000	32.31011	55%
Simulation 4	0.9	1000	83.60323	40%
Simulation 5	1	1000	135.21421	10%

8.4.4 Momentum Effect

The following experiment will use the best Learn Rate value that is obtained from previous experiments and uses different Momentum grades (0.1; 0.2; 0.5; 0.7; 0.91) to analyze the best result based on the Momentum grade. Another parameter, such as initial reach, is still the same as with the previous experiment. The following is some simulation done to Momentum grade change.

From the simulation, the conclusion can be obtained that JSB has a better learning trend in choosing symmetrical reach and, to the contrary, is not symmetrical. From Simulation 7 and Simulation 8, it seems that the smaller magnitude (-2 to +2), JSB also has trend 6 to learn better. Therefore, in the next experiment, the grade reach will be used.

From the above simulations it can be concluded that the best Momentum rate is obtained in Simulation 3 with Momentum grade = 0.5. In Simulation 3 obtained Mark Level successful grade = 100 percent in epoch 230 that shows JSB can learn better in the condition. Therefore, Learn Rate grade combination and Momentum can temporarily be manual to the next analysis.

8.4.5 Initial Value Reach Effect

The following is an experiment to analyze initial value reach effect to JSB. In this experiment JSB will be used, with the best characteristic obtained from the previous experiment, which is Learn rates=0.3 and Momentum grade = 0.5. Following are some simulations towards initial value reach difference.

8.4.6 Regional Composition Effects towards JSB Sample

All the experiments obtained a good JSB characteristic, with Learn Rates value parameter = 0.3; Momentum = 0.5, and initial value reach = -2 to +2. The experiment has proved that the letter pattern introduction method and hand-written numbers made by the writer can be "learned" well by JSB. This

TABLE 8.5

Momentum Effect Experiment Result

Experiment	Momentum	Epoch	Error Level	Match Level
Simulation 1	0.1	1000	26.21457	92%
Simulation 2	0.2	1000	47.13567	62%
Simulation 3	0.5	1000	16.63731	100%
Simulation 4	0.7	1000	34.43421	70%
Simulation 5	0.9	1000	27.47765	77%
Simulation 6	1	1000	53.37567	65%

TABLE 8.6

Reach Effect Experiment Result

Experiment	Reach	Epoch	Error Level	Match Level
Sim 1 (A)	0 s/d +2	1000	34.33457	0%
Sim 2 (A)	0 s/d -5	1000	57.24567	0%
Sim 3 (A)	-5 s/d 0	1000	23.76531	87%
Sim 4 (A)	-2 s/d 0	1000	44.47621	54%
Sim 5 (A)	-5 s/d 2	1000	38.45625	71%
Sim 6 (A)	-2 s/d -5	1000	61.39867	31%
Sim 7 (A)	-2 s/d -2	230	36.37857	100%
Sim 8 (A)	-3 s/d -5	100	58.29867	100%

TABLE 8.7

Region Composition Effect Experiment Result

Experiment	Region	Epoch	Error Level	Match Level
Simulasi 1	5*4	230	23.71231	100%
Simulasi 2	4*5	1000	51.65625	71%
Simulasi 3	4*4	1000	50.83221	43%
Simulasi 4	4*3	1000	48.67028	54%
Simulasi 5	3*4	1000	38.97013	91%
Simulasi 6	3*3	1000	53.22901	91%

is recognized by the reach of Match Level value = 100 percent with the characteristic meant. Otherwise, to complete the experiment and analysis to the methods, the writer carries out an experiment that is Region composition effect towards JSB characteristic and final result that will be obtained. The result of several simulations can be seen in Table 8.6. From the six simulations it can be seen that Simulation 1 gives the best learning result. Therefore, it can be concluded that temporary Region 5*4 composition is the best.

8.5 Conclusion

Based on process and test it can be concluded that:

1. The introduction of a three-dimensional image pattern in this study can be implemented with a counter-propagation neural network.
2. The results obtained are based on the analysis of experimental data that has been carried out for the level value with the best result with this method being 0.5.

3. The best momentum value from this method is 0.6.

4. The method that has been used directly is proven to be able to study and identify the pattern using artificial neural networks.

5. Based on the results of the research that has been done, it is stated that the results of the three-dimensional image analysis are very accurate and significant.

8.6 Exercises

1. Describe the term propagation neural network.

2. Give an example of the application of artificial neural networks in the health sector.

3. Describe the steps of the algorithm in neural networks.

4. Give your explanation of image processing.

5. Give an example of a case study about image processing.

6. How do you calculate the process energy function in an artificial neural network?

7. How do you calculate the weight and weight in the artificial neural network?

8. How do you calculate the numeric data extraction algorithm?

9. How do you calculate the numerical data normalization results?

10. Give an example of a simulation of the problem solved using an artificial neural network.

References

[1] Alex Berson and Stephen J. Smith, *Data Warehousing, Data Mining and Olap.* New York: McGraw Hill, 2000.

[2] Foussete, Laurence, *Fundamental of Neural Networks: Architecture, Algorithm, and Applications*, Hoboken, NJ: Prentice Hall, 2017.

[3] Fu, Limin, 2016, *Neural Networks in Computer Intelligence*, New York; McGraw Hill.

[4] Hidayatno, A., Isnanto, R. R., and Buana, D. K. W., 2018, Signature Identification Using Backpropagation Artificial Neural Networks, http://core.ac.uk/downl oad/pdf/11728578.pd f, accessed on 6 November 2015.

[5]   Hayatunnufus, A., Andrizal., and Yendri, D., 2013, Signature Detection and Verification Using the Spatial Image Domain Method, http://repository. unand.ac.id/18733/ 1 / Jurnal% 20Annisa% 20Hayatunnuf us.pdf, accessed on 6 November 2015.

[6]   Kosko, Bart, 2016, *Neural Networks and Fuzzy Systems: A Dynamic System Approach to Machine Intelligence*, New Delhi Prentice Hall India.

[7]   Kullcorni, Arun D., 2016, *Artificial Neural Networks for Image Understanding*. New York: Van Nostrand Reinhold.

[8]   Purnomo, M. H., and Muntasa, Arif., 2017. *The Concept of Digital Image Processing and Feature Extraction*. Yogyakarta: Graha Science.

[9]   Putra, Darma, 2017. Digital Image Processing. Yogyakarta: Andi Offset. Utomo, E. B., 2017, Face Recognition of a Veiled Woman Using the 2DPCA and K-Nearest Neighbor Methods, http://eprints.dinus.ac.id/16475/1/ jurnal_15 403.pdf, accessed on December 10, 2017.

[10] Yang, J., Zhang, D., Frangi, A. F., and Yang, J. Y. 2016. Three-Dimensional PCA: A New Approach to Face Representation Based on Appearance and Recognition, www.dtic.upf.edu, accessed on 10 December 2016.

9

The Evolution of Emerging Market (EM) Sovereign CDS Spreads during COVID-19

Nadir Oumayma and Daoui Driss

CONTENTS

9.1 Introduction

The COVID pandemic triggered an economic crisis with vicious violence. Regional and national closures led to disruptions in global value, chain stores and home orders squeezed consumer demand and, as a result, business and personal incomes and oil prices plummeted. In fact, this price drop occurred as the United States and Europe experienced their first waves of COVID, while many emerging countries were in the early stages of their infection curves. As a result, for some of the world's emerging and developing countries, the

DOI: 10.1201/9781003126218-9

effects of the pandemic manifested themselves first in foreign trade and only then in domestic health issues. In particular, commodity-dependent emerging markets went through a particularly difficult first half of 2020. They were first hit by a surge of oil and other commodities and also with the flow of COVID infections. In this chapter, we contribute to the discussion of the relative importance of global factors. Specifically, we compare the importance of dominant market factors in relation to the dynamics of COVID-19 and policy responses to explain the evolution of emerging market sovereign Credit Default Swap (CDS) spreads in the first half of 2020. In doing so, we unravel the effects of the global economic and financial turbulence due to the effects of lower oil prices on emerging markets. The analysis focuses on daily emerging market sovereign CDS spreads, comparing the impact of dominant market factors with the COVID-19 dynamics, the European Central Bank (ECB) and the Federal Reserve Bank (FED) policies, and countries' fiscal policies regarding sovereign spreads and fiscal adjustments to collapsing demand. We adopt a multistage econometric approach, drawing on a dataset from an international panel of MEs. In the first step, we estimate a multifactor model for CDS changes spanning several years before the emergence of COVID-19. Specifically, we use the model from January 2014 to June 2019, the "pre-COVID". In the second step, we use the estimated coefficients from the first step to extrapolate the model's implicit changes in CDS spreads from July 2019 to June 2020. This facilitates the statistical derivation of the "COVID residual", that is, the difference between the actual CDS adjustment and the change implied by the model. In the second step, we also explain the implicit model and the actual CDS changes by COVID-related factors.

In other words, we examine the residuals and explain them by a panel analysis that uses COVID-related factors in three different areas: epidemiological, economic and political factors.

9.2   Recent Literature

This chapter contributes to the literature that attempts to understand the relative importance of global and country-specific factors. We do not attempt to provide a general overview of the previous scholarship but, instead, refer readers to August 2014 for a recent listing of related studies. Intuitively, one might expect that the pricing of sovereign risk would be determined by country-specific factors, but there is evidence that some of the variation in sovereign CDS spreads is determined by global variables and unrelated to the country-specific context.

This is particularly true for high trading frequencies, although, as we shall see, this is not the case for all periods. Ultimately, it is an empirical question whether global factors, country-specific factors, or both are together responsible for the dynamics of CDS spread. To get an overview, we review the main academic findings from three different camps: the "pro-global", the "pro-local", and the agnostic camp and focus specifically on exchanges dealing with CDS spread markets. This puts our results into perspective and demonstrates the practical use of our method.

9.2.1 Local Risk Factors

Using data from 2003 to 2012, Ertugrul and Ozturk [3] study the relationship between CDS spreads and financial market indicators for bonds, equities, and market currencies for selected EU countries. Their results suggest that CDS spreads have a cointegrating relationship with other financial market indicators for the entire sample.

Another conclusion that deserves special attention is that, in the long run, the CDS spread is negatively related to the uncertainties in the CDS market. They argue that this negative relationship indicates low liquidity under high uncertainty, which drives down CDS prices. The variation over time in the effects of each variable on the CDS spread is consistent with the results obtained from the cointegration analyses.

Covering the so-called "taper-tantrum" episode of 2013 and other EM episodes [1] assesses the importance of the use of money in debt management. Cross-country regressions lead them to the following results: (1) Emerging economies with relatively better economic fundamentals experienced less deterioration in financial markets during the 2013 market meltdown episode. (2) Differentiation between emerging financial markets: Markets were established relatively early and persisted throughout this episode. (3) During the temper tantrum, while holding that neither macroeconomic variables nor country ratings significantly explain the spread of CDS changes. Second, measures of United States bonds, equity, and CDX High-Yield yields are the main drivers of CDS spread changes. Finally, their analysis suggests that CDS spreads are more strongly influenced by international spillovers in times of market stress than in normal times, which leaves some room for country–specific factors.

9.2.2 Global and Local Risk Factors

While the scholarships of the "pro-global" and "pro-local" camps generally recognize each other, they do so more or less casually. However, based on the intuition that financial asset prices are determined by country-specific fundamentals, Remolona and colleagues [14] decompose monthly data on risk appetite in emerging markets at five years. CDSs are spread over the period 2002 to mid-2006 into a market-based measure of expected loss and a risk premium. They analyze how each of these two elements relates to country-specific risk measures and the global risk aversion/risk appetite measure. Fundamental variables include inflation, industrial production, consensus forecasts of GDP growth, and currency reserves. Indicators of global risk aversion are as follows: Tsatsaronis and Karamptatos [16], the effective risk appetite indicator, the VIX and a risk-tolerance index by J.P. Morgan Chase. They find empirical evidence that global risk aversion is the main determinant of sovereign risk, whereas country-specific fundamentals and market liquidity are more important for risk. The two components therefore behave differently.

9.3   Overview of Emerging Markets And COVID-19

In contrast to previous crises, the response of emerging markets to the impact of COVID-19 has been decisive: With the exception of Saudi Arabia, all member states put in place fiscal and non-fiscal stimulus packages. While the stimulus packages in emerging market economies may seem impressive at first glance, Alberola and colleagues [2] point out that they are not that large compared to advanced economies. In fact, fiscal measures in advanced economies reached 8.3 percent of GDP, 6.6 percentage points (pp) more than following the GFC, while in emerging markets they amounted to only 2 percent of GDP, even less than in the emerging markets of the last financial crisis. The contrast is most striking for credit guarantees: 6.6 percent of GDP in advanced economies and only 0.4 percent in emerging markets. The gap is narrower for financing facilities: 4 percent of GDP in advanced economies compared with 1.3 percent in emerging markets. The difference in fiscal stimulus between advanced and emerging markets may indicate a lack of space for the latter. But it is likely that fiscal constraints account for only part of the difference; another possible reason for the reduction in the fiscal stimulus packages is the lack of space in the advanced and emerging markets could be the difference in the prevalence of the pandemic: COVID-19 affected advanced economies earlier and more strongly than emerging markets, with the exception of some Asian emerging markets.

Contrary to budgetary constraints, emerging markets seem to have had some leeway in terms of conventional monetary policy over advanced economies. In fact, emerging markets have been able to take advantage of greater leeway to reduce policy rates than their advanced counterparts. At the beginning of 2020, policy rates in emerging markets averaged 4.9 percent (excluding Argentina), while the average policy rate for advanced economies was 0.4 percent.

Since then, member states have cut policy rates by about 114 basis points (excluding Argentina), compared with 40 basis points for advanced economies. However, rate cuts alone are not a panacea for emerging markets. Especially for oil-exporting countries (except for Mexico, which largely covers oil revenues), rate cuts had to go hand in hand with foreign exchange interventions.

9.4 Methodology

In this section, our main objective is to examine whether and to what extent COVID-related developments and the associated policy responses of countries, central banks, and the IMF have influenced the pricing of sovereign risk in emerging markets. In this context, we ask whether and to what extent the dominant factors and dynamics related to COVID have influenced sovereign risk pricing by the pandemic. We propose a two-step econometric analysis that takes advantage of a daily data set of cross-national panels.

In the first step, we estimate a dynamic heterogeneous multi-factor model for changes in five-year CDS spreads over the period from January 1, 2014 to June 30, 2019. Then, using the estimated parameters of the model, we apply a synthetic control-type procedure to extrapolate the model – implicit change in CDS spreads, given by the realized values of the factors – from July 1, 2019 to June 30, 2020.

This approach allows us to calculate the "COVID residual", that is, the difference between the realized CDS adjustment to the variation implied by the model at both the individual country level and the aggregate Emerging Markets (EM) level, during the pandemic period.

In a second step, focusing specifically on the 2020 pandemic period, we examine whether daily deaths due to COVIDs, announced policy responses, or other country fundamentals contribute to explaining the variation in the COVID residual.

We are taking several steps to get as close as possible to a causal interpretation of our results.

First, we control for a series of alternative explanatory variables that could distort our results, and second, our results are not influenced by time factors in our example, such as the advantage of certain countries being democratic/non-democratic or having a specific currency.

9.4.1 Stage 1 Estimate, January 2014–June 2019 (Table 9.1)

First, we estimate and evaluate a heterogeneous multi-factor model. Our empirical analysis uses daily data for 30 emerging market sovereigns over a

TABLE 9.1

First-phase Regression Results Estimated for the Period January 2014 to June 2019

Dependent variable ΔCDSit	Δcds$_{it-1}$	ΔGCDS$_t$	ΔRCDS$_t$	R-Carré	Out of sample R-Square: July 1st, 2019 to June 15, 2020
Germany	−0.395***	0.322***	0.820***	0.22	0.17
France	−0.223***	0.117	1.156***	0.17	0.19
Greece	−0.019	0.083	0.395***	0.04	0.32
Ireland	−0.050**	0.179***	0.905***	0.28	0.26
Belgium	−0.368***	0.300**	0.578***	0.16	0.17
Spain	−0.337***	0.430***	1.833***	0.33	0.17
Netherlands	−0.212***	0.235***	0.566***	0.15	0.22
Austria	−0.269***	−0.097	0.955***	0.17	0.12
Cyprus	−0.125***	0.161*	0.149**	0.03	0.20
Estonia	−0.222***	−0.156***	0.167***	0.09	0.07
Italy	0.021	0.305***	1.470***	0.37	0.16
Latvia	−0.034	0.407***	0.158***	0.09	0.01
Lithuania	−0.093***	0.275***	0.189***	0.10	0.01
Portugal	−0.052**	0.407***	1.229***	0.30	0.55
Slovenia	−0.127***	0.170***	0.200***	0.09	0.01
Slovak Republic	−0.168***	0.305***	0.173***	0.16	0.01
Finland	−0.175***	0.209***	0.459***	0.18	0.19

NB: Country-specific time series regression estimates from equation 1. Dependent is the change in the daily CDS spread. ***, **, * correspond to 1, 5, and 10 percent significance, respectively. Out-of-sample (pseudo) R-squared reports the percentage change in the real Δcdsit explained by the model's implied values over the out-of-sample estimation period, July to December 2019. Number of daily observations per country, T, equal to 1.432.

period of 6.5 years, from January 10, 2014 to June 30, 2020. We selected 30 MEs based on their undesirability and the availability of data for the dependent variable where undesirability is defined by the representation of a country in the benchmark index for ME sovereigns, the J.P. Morgan Emerging Markets Bond Index.

We chose this specific period because it begins after the structural break in the temper tantrum and to have sufficient data to calibrate and test the model under normal and COVID-times conditions.

9.4.1.1 Data

We use the following data:

- Sovereign CDS (CDS spread). We use daily 5-year CDS spreads reported by Eikon Refinitiv and convert the levels into daily log changes.
- Gross Domestic Product (GDP). We use GDP data in current $ reported by the World Bank.

9.4.1.2 Specification

First, we estimate a dynamic factor model on the data from the pre-COVID period of the following form:

$$\Delta cdsi,t = \alpha i + \phi i \Delta cdsi,t-1 + \beta i1 \Delta GCDSt + \beta i2 \Delta GCDS\acute{\imath},j,t+\varepsilon i,t \qquad (1)$$

January $1.2014 \leq t < $ July 1.2019

$$\text{And } \Delta CDSi,t = In\ (CDSi,t)/(CDSi,t-1)$$

Our result variable is the daily variation of the logarithm of the dispersion of CDS in the country i. On the right, we include the lagged dependent variable and two factors. A global factor, $\Delta GDCSt$ and a regional EM factor $\Delta RDCSi'$, j, t. The global factor is constructed as the GDP-weighted cross-sectional average of the daily changes in log CDS over a reference group of 20 advanced economies: the United States, Japan, and 18-euro area member states. It therefore captures the common component of fluctuations in sovereign risk at the global level. The regional factor is constructed slightly differently. It is the GDP-weighted cross-sectional average of the daily variations in the logarithm of CDSs relative to the reference sovereigns of a country in its region. In other words, the 30 emerging markets were first classified into seven reference groups based on their geographical proximity and dependence on oil exports.

We estimate the model over the pre-COVID period from January 1, 2014 to June 30, 2019.

Instead of estimating the model to the end of 2019, we chose this window because it extends from July to December 2019 to validate the out-of-sample accuracy of our model before the COVID shock in 2020. Finally, the COVID residual is defined as follows:

$$\Delta cr_{i,t} = \Delta cds_{it-} [\check{\alpha}_i + \phi_i \Delta cds_{i,t-1} + \beta_{i1} \Delta GCDS_t + \beta_{i2} \Delta GCDS_{i,j,t}] \qquad (2)$$

By simply comparing the realized change in log CDS to the expected value of the model, and taking into account the actual realization of the factors and the lagged change in log CDS.

9.4.1.3 Exposure to global and regional risks and fiscal fundamentals in emerging markets

Regional and global betas are positively, but not significantly, associated with the size of COVID-related fiscal stimulus in emerging markets. This leaves room for two interpretations: On the one hand, since the coefficients are not statistically significant, it cannot be excluded that systematically riskier countries (higher regional betas) issued less stimulus/GDP due to considerations of lack of fiscal space. On the other hand, if Saudi Arabia, which is an outlier, were removed from the sample, the coefficients would become potentially significant. This, in turn, could mean that although countries with high betas would generally find it difficult to engage in deficit spending, the seriousness of the situation, coupled with a low interest rate environment, could mean that countries that would otherwise be perceived as risky could engage in economic stimulus.

9.4.2   Second Estimation Phase, January–June 2020

For the second stage of estimation, we separate the out-of-sample period from January to June 2020 into three COVID sub-periods:

- January–February 2020 (start of COVID)
- March 2020 (peak COVID)
- April–June 2020 (end of COVID)

9.4.2.1   Data

- Infections. We use daily infections reported by country by the Johns Hopkins University Center for Systems Science and Engineering (JHU CSSE). Figures include confirmed and probable when reported.
- Deaths. We use the daily deaths by country reported by JHU CSSE. Figures include confirmed and probable where it was reported.
- Mobility. We use daily routing requests by country reported by Apple (driving; walking; public transit).
- Index of stringency. We use the Oxford COVID-19 Government Response Tracker as a measure that records the stringency of "lock-in style" policies that primarily restrict people's behavior.
- Daily fiscal and monetary policy announcements. This information was collected for individual countries, for the Central Bank, and for the Federal Reserve. These columns indicate whether an action or proposal was made by a given nation/institution on a specific date in the but do not control the size or number of fonts on a given day. A line is coded "1" if the date corresponds to the announcement of at least one political key. Except for the Federal Reserve (whose major announcements were related to interest rate cuts and budget spending), we have restricted our analysis of key fiscal policies to those that provided "millions" or "billions" of local currency spending units.

The extended list of sources that used to aggregate these data can be found in the online appendix.

- External debt. We use the total stock of external and private sector debt as a percentage of GDP as reported by the World Bank. The data are annual.
- Debt owed to China. We use the estimate of the total stock of external debt owed to China in current US dollars as a proportion of the debtor's GDP as reported by Horn and colleagues [8].

9.4.2.2 Specification

We first estimate a panel model examining the relationship between the COVID residual, defined as the difference between the realized values of the daily variation in the spread of CDS and the implied values of the dynamic factor model (1), and a set of COVID-related variables from three different domains: pandemic, economy, and policy measures. The second step of the model specification is as follows:

$$\Delta cr_{it} = \vartheta_i + \Lambda_t + \Theta X_{it}^{COVID} + y X_{i,t}^{economy} + \eta X_{i,t}^{policy} + \varepsilon_{i,t} \tag{3}$$

The variables specific to a pandemic: mortality outcomes, in which we include daily new mortality (per 1 million population), daily growth rate of new mortality, total mortality rate (per 1 million population), and total mortality growth rate, measure of daily mobility in terms (reported by Apple), and daily growth rates of policy stringency indices (constructed by OxCGRT). Lower levels of mobility or stricter non-pharmaceutical government interventions may signal greater economic contraction, which may increase the burden of debt financing and thus had an impact on the price of debt during the COVID-19 pandemic.

Economic variables include the effect of oil prices on revenues, request for information (RFI) announcements, sovereign fund reserves, external debt ratios, and debt to China ratios.

It is interesting to note that all three policy measures, that is, the FED, the ECB, and country-specific fiscal policies, have statistically significant associations with COVID residuals. Specifically, the interaction of fiscal policy with the level of external debt is positive, indicating that countries that have increased their debt burdens through stimulus measures, or countries that already have relatively high debt burdens were likely to see a larger gap in the dynamics of CDS spread. More (hidden) debt to China is also positively associated with CDS residuals, although not statistically significantly so, in all specifications. However, the interaction of debt to China with the fiscal policy dummy is again negatively correlated with CDS spreads and statistically significantly, as is the interaction of (non-hidden) external debt with the fiscal policy dummy. This suggests some investor indifference to the sustainability of the announced stimuli, which can probably be explained in part by the fact that many financial market players are not aware of the debt that countries owe to China.

Finally, the effect of oil revenues is statistically highly significant in specifications 3 and 4, indicating that oil exporters experience a relative compression of COVID residues compared to oil importers. A noteworthy finding is the F-statistic in the first specification. Since this statistic is not significant, we cannot reject the fact that the group of COVID variables is jointly insignificant. This means that the COVID residuals are not due to COVID-specific risks but rather to traditional drivers of sovereign debt pricing, such as the margin of fiscal maneuvering, the effects of oil revenues and global factors such as the and European monetary policy, because Specification 4 contains jointly significant variables.

9.4.2.3 Residue Review, March 2020

Next, we use a modified specification to compare the explanatory power of the predictions of the factor model and the variables related to COVID. We do this by treating the logarithmic changes in CDS spreads as the outcome variable, while increasing the panel regression with the model's implied values of (1) on the right-hand side and the COVID-related variables as in (3)

$$\Delta cds_{i,t} = \vartheta_i + \Lambda_t + \Gamma\, \Delta cds_{i,t} + \Theta X_{it}^{COVID} + y X_{i,t}^{economy} + \eta X_{i,t}^{policy} + \varepsilon_{i,t} \tag{4}$$

With

$$\Delta cds_{it} = \check{\alpha}_i + \phi_i \Delta cds_{i,t-1} + \beta_{i1}\Delta GCDS_t + \beta_{i2}\Delta GCDS_{i,j,t}$$

These are the implied values of the daily changes in the CDS spread generated by the factor model:

(1) The fixed effects are again designated by ϑi and λt, representing the fixed effects by country and period respectively.

Essentially, in (4), we separate the two components that make up the COVID residual. In this way, it becomes clear that the regression in (3) with the COVID residual as an outcome variable is equivalent to the restricted regression in (4) when $\Gamma = 1$. (4) relaxes this assumption implicit in (3) and allows for a richer analysis.

9.5 Results

First, after including the COVID-related variables, the coefficient of the model's implied values remain statistically significant and keep their positive sign, implying that the model's implied values still track the changes in CDS spread achieved during the COVID-19 pandemic, giving merit to our model selection in the first instance.

Second, both the daily new mortality and the growth rate of new mortality are positively and significantly correlated with the daily changes in CDS spread over the specification set. Consistent with the evolution of the time series of COVID residuals shown in the previous section (Figure 9.1), countries that had higher levels of new daily mortality rates or higher new mortality growth rates were likely to experience a more severe daily spread of CDS changes.

Third, announcements of country-specific budget measures to address the pandemic appear significantly associated with daily CDS changes. Countries that announced fiscal responses and thus increased debt burdens were more likely to experience daily variations in CDS spreads.

It is important to note that our results show that factors specific to COVID explain most of the variation in CDS spread dynamics during the COVID-19 pandemic period. In particular, mortality results, including new daily mortality dynamics, explain about 6 percent of the variation, with COVID-specific policy announcements adding another 3 percent. In contrast, dynamic factor model predictions explain only about 2 percent of the variation over this period, implying that the explanatory power of COVID-specific factors is almost five times greater than that of regional and global factors that did a good job for 18 times the normal range.

The high-frequency regression parameter of the country panel may lead to low overall explanatory power even after controlling for COVID-specific factors in addition to regional and global factors.

In Figure 9.2, we present the overall (average) dynamics of CDS spread for the euro zone countries during the pandemic by plotting the overall realized and implied values of the model. Surprisingly, the global values implicit in our regressions (which consider COVID-specific factors) almost perfectly follow the realized values, so that their lines coincide with each other.

We can see this by plotting the realized aggregate values, the model implied values in column (1) of Table 9.3 and the model implied values in column (4) of Table 9.3. Surprisingly, the global values implied by the model in equation (4), which for COVID -specific factors, almost perfectly traces the realized values, so that their rows coincide with each other. We therefore conclude that COVID-specific factors play an important role in explaining the divergent dynamics of CDS spread during the pandemic and should not be ignored in the pricing of debt in the euro area during this period.

A potential limitation of the second-step analysis, which may contribute to the overall low R-squared values of the model, is the omission of important variables due to data limitations.

For example, credit tenure could play a key role in the eligibility of assets for purchase by the ECB, but we have no control over this. Another important

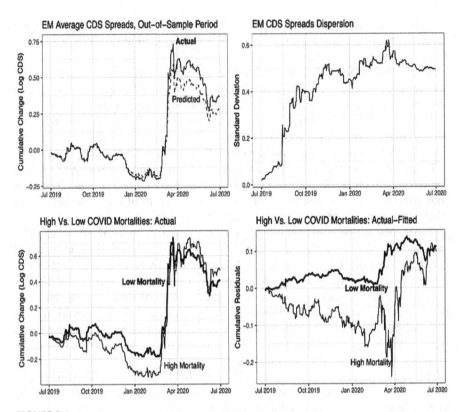

FIGURE 9.1

Emerging market sovereign CDS from July 2019 to June 2020.

Note: The top left panel of the figure plots the actual changes in the mean cumulative ME (log) CDS spread over the period COVID-19 (solid) compared to those implied by the factor model (dashes). First, it should be noted that the factor model does a good job of predicting changes in the CDS spread to the end of 2019. However, from 2020 onwards, the actual and predicted series start to diverge. The gap between actual and predicted only begins to narrow in March 2020, when governments and central banks around the world announced economic stimulus packages. The top right panel shows the cross-country dispersion of CDS spreads over the same period. While the dispersion increased in the second half of 2019, it increased in March 2020, highlighting the emergence of country-specific exposure. The lower panels compare the five countries with the highest mortality rates to the five countries with the lowest mortality rates in April. The lower left panel suggests that the evolution of actual CDS spreads has been similar for high mortality and low mortality countries. Nevertheless, the lower right panel indicates that the high-mortality countries initially experienced larger fluctuations in their CDS spreads, but that the gap between actual and expected values became similar to that of the low-mortality countries in July.

Source*: Hevia, C, P A Neumeyer (2020), "A perfect storm: COVID-19 in emerging economies", VoxEU.org, 21 April.

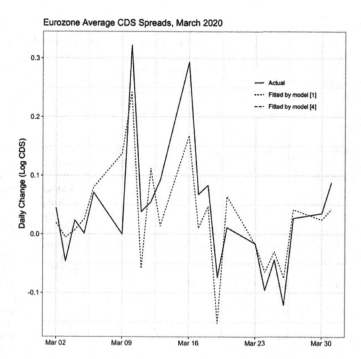

FIGURE 9.2

Risks and factors related to COVIDs considered for average CDS in the euro zone.

*Note: Solid lines reflect average daily changes in euro zone CDS spreads. Dotted lines reflect the average expected changes in euro area CDS spreads, as specified in [1] and [4] in Table 9.2, respectively.

issue is the interpretation of global and regional factors as reflecting fundamental proxies. If COVID-19 is a global shock, then our interpretation of the global factor may not be accurate, as the variation in the global factor may incorporate both fundamental and pandemic information. This is a special case of the more general problem of identifying idiosyncratic or country-specific variations in macroeconomic and financial research, a common dilemma when exploring the extent of financial contagion [17].

9.6 Conclusion

We investigate the role that country-specific and global variables have in determining emerging market sovereign CDS during the pandemic crisis. We use a two-stage econometric model approach. First, a heterogeneous factor

TABLE 9.2

Residual COVID Sovereign Deviations, Reduced Sample, Pandemic Period March 2020.

Sovereign Variances Residual COVID

	(1)	(2)	(3)
New mortality rate	0.0096**	0.0102*	0.0108*
	(0.0043)	(0.0055)	(0.0056)
New growth in mortality rate	0.0033***	0.0035***	0.0033***
	(0.0008)	(0.0010)	(0.0010)
Total mortality rate	–0.0004	–0.0004	–0.0004
	(0.0006)	(0.0008)	(0.0008)
Growth in total mortality rate	–0.0456	–0.0510	–0.0510
	(0.0433)	(0.0482)	(0.0492)
Mobility		0.0004	0.0003
		(0.0006)	(0.0006)
SI Growth		0.0093	0.0141
		(0.0553)	(0.0567)
Country-by-country budget policy announcement			0.0394*
			(0.0219)
Announcement on EU tax policy			–0.0065
			(0.0647)
ECB policy announcement			0.0323
			(0.0696)
Fed policy announcement			0.0414
			(0.0747)
Fixed Effects	Y	Y	Y
Observations	156	149	149
R2	0.0416	0.0423	0.0600
F Statistic	1.2362	0.7883	0.6569

Note:
Pandemic Sample: Data as of March 2020; Residual COVID: The difference between the actual adjustment of the CDS and the change implied by the model, both at the level of individual countries and at the level of the aggregated EA (Embryonic Zebrafish) over the pandemic period. *, **, *** correspond to an importance of 10%, 5% and 1%, respectively. Robust HAC standard errors, grouped by country. EFPs of time and country.

model is used to "form" a model that predicts daily changes in CDS spreads based on previous changes and overall CDS dynamics. The method is useful for addressing various data issues that are particularly relevant in emerging markets, such as data unavailability and differences in data frequency.

The model makes relatively accurate forecasts both in-sample (January 2014–June 2019) and out-of-sample (July 2019–December 2019). In 2020, forecasts lose precision and residuals increase. Given that COVID is a global pandemic and that the spread of CDS in countries around the world has been affected, we consider the increase in residuals as a sign that country-specific variables, in particular drive CDS spreads in times of crisis. The residuals

TABLE 9.3

Panel Analysis of Daily Change in CDS spreads, Pandemic Sample

Dependent variable / Daily change in CDS Gap	(1)	(2)	(3)	(4)
Daily change in CDS spreads	0.3689**	−0.4345**	−0.4707**	−0.5135***
	(0.1797)	(0.1912)	(0.1954)	(0.1856)
New mortality rate		0.0086**	0.0087*	0.0094*
		(0.0034)	(0.0047)	(0.0050)
New growth in mortality rate		0.0038***	0.0038***	0.0029***
		(0.0008)	(0.0011)	(0.0010)
Total mortality rate		−0.0005	−0.0005	−0.0004
		(0.0004)	(0.0006)	(0.0006)
Growth in total mortality rate		0.0169	−0.0151	−0.0135
		(0.0407)	(0.0478)	(0.0473)
Mobility			0.00002	−0.0001
			(0.0008)	(0.0007)
SI growth			0.0044	0.0069
			(0.0464)	(0.0407)
Country-by-country budget policy announcement				0.0475**
				(0.0218)
Announcement on EU tax policy				0.0138
				(0.0584)
ECB policy announcement				−0.0261
				(0.0678)
Fed policy announcement				−0.0509
				(0.0558)
Effects fixes	Y	Y	Y	Y
Observations	374	156	149	149
R2	0.0206	0.0812	0.0866	0.1184
F Statistic	7.0406***	1.9975*	1.4351	1.2448

Note:
Pandemic Sample: Data after March 2020; COVID Residual: The difference between the actual CDS adjustment and the change implied by the model, both at the individual country level and the aggregate EA levels during the pandemic period. *, **, *** correspond to 10%, 5% and 1 respectively. Robust HAC standard errors, grouped by country. FEs by time and country.

are particularly significant at the time of peak COVID in March 2020, which corroborates several papers in the literature that focus on the nature of the time variation in the relationship between autonomous CDS spreads and explanatory variables. Second, regressions of residuals on fundamentals suggest that COVID mortality and infections are not as important as the variables to explain CDS spreads the capture of fiscal space, economic activity, FED and ECB actions, and oil price changes.

There are two possible interpretations of the relative insignificance of COVID mortality and infections for changes in CDS spread. On the one hand, international investors may view COVID mortality and infection data as noisy and unreliable and not pay too much attention to them when making their investment decisions. On the other hand, international investors may be aware that COVID is a pandemic that may ultimately affect all countries to roughly the same degree in terms of health impacts and, therefore, focus less on infections and mortality and more on economic performance and stability.

An interesting empirical corollary is the finding that while the external debt and GDP levels of emerging countries are not statistically significant in explaining the COVID residuals, the (hidden) debt to China is statistically significant.

References

[1] Ahmed, S., Coulibaly, B., and Zlate, A. (2017). International financial spillovers to emerging market economies: How important are economic fundamentals? *Journal of International Money & Finance*, 76, 133–152.

[2] Alberola-Ila, E., Arslan, Y., Cheng, G., and Moessner, R. (2020). The fiscal response to the Covid-19 crisis in advanced and emerging market economies (No. 23). Bank for International Settlements.

[3] Ertugrul, H. M., and Ozturk, H. (2013). The drivers of credit default swap prices: Evidence from selected emerging market countries.

[4] *Emerging Markets Finance and Trade*, 49(sup5), 228–249.

[5] Fender, I., Hayo, B., and Neuenkirch, M. (2012). Daily pricing of emerging market sovereign CDS before and during the global financial crisis. *Journal of Banking & Finance*, 36(10), 2786–2794.

[6] Forbes, K. J., and F. E. Warnock (2012), Capital flow waves: Surges, stops, flight, and retrenchment, *Journal of International Economics* 88: 235–251.

[7] Heller, P. (2005), *Understanding Fiscal Space*. IMF Policy Discussion Paper 05/4. International Monetary Fund, Washington, DC.

[8] Horn, Sebastian, Carmen M. Reinhart, and Christoph Trebesch. (2019), *China's Overseas Lending*. NBER Working Paper No. 26050. The data is yearly. (HRT _ China Debt Stock Database_update_April2020.xlsx).

[9] Ismailescu, I., and Kazemi, H. (2010), The reaction of emerging market credit default swap spreads to sovereign credit rating changes. *Journal of Banking & Finance*, 34(12), 2861–2873.

[10] Jinjarak, Yothin, et al. *Accounting for Global COVID-19 Diffusion Patterns, January-April 2020*. No. w27185. National Bureau of Economic Research, 2020.

[11] Jinjarak, Yothin, et al. (2020). *Pandemic Shocks and Fiscal-Monetary Policies in the Eurozone: COVID-19 Dominance During January–June 2020*. NBER Working Paper No. 27451.

[12] Kocsis, Z., and Monostori, Z. (2016). The role of country-specific fundamentals in sovereign CDS spreads: Eastern European experiences. *Emerging Markets Review*, 27, 140–168.

[13] Panjwani, Ahyan, Ross, Chase. (2020), A Long Way to Go for Emerging Markets. Retrieved from: https://som.yale.edu/blog/long-way-to-go-for-emerging-markets.

[14] Remolona, E. M., Scatigna, M., and Wu, E. (2008). The dynamic pricing of sovereign risk in emerging markets: Fundamentals and risk aversion. *The Journal of Fixed Income*, 17(4), 57–71.

[15] Rey, Hélène, *Dilemma not Trilemma: The global financial cycle and monetary policy independence*. Federal Reserve Bank of Kansas City working paper (2013). 30.

[16] Tarashev, N., Tsatsaronis, K., and Karampatos, D. (2003). Investors' attitude towards risk: what can we learn from options. *BIS Quarterly Review*, 6, 57–65.

[17] Bekaert et al. (2014). Global crises and equity market contagion/working Paper 17121. http://www.nber.org/papers/w17121

10

Prediction of COVID-19 Data Using Business Intelligence Tools

Sara El Habbari, Mhamed-Amine Soumiaa, and Mohamed Mansouri

CONTENTS

10.1 Introduction

Predictive analyses use techniques such as machine learning and deep learning, which are artificial intelligence (AI) technologies, to predict what is likely to happen. They will never be able to predict the future, but they can examine existing data and establish a probable outcome. This type of analysis is important in business intelligence (BI) because it can help to process large data volumes in real time, have a vision of what is likely to happen in the future, and anticipate the appropriate actions to take to increase profits and avoid crises.

Despite the importance of predictive analyses in BI, their usage rate remains low compared to other types of analyses, such as descriptive and diagnostic ones. According to Dresner Advisory Services, survey results reported in the

DOI: 10.1201/9781003126218-10

American economic magazine *Forbes* [1] state that companies focus more on historical data to better understand past events. This observation is confirmed by the results of an IDC France study [2], which reveals a rate of only 9 percent of French enterprises that have used predictive solutions, even though they are a priority for 26 percent of business departments. According to the same study, this low rate can be explained by organizational, technical, and human obstacles, especially in large company structures.

Currently, several scientific articles talk about predictive analyses, but very few link them to BI. The research studies found along those lines ([3] and [4]) uses BI tools to mainly understand the data (correlations, frequencies, etc.) and then exploits the results on predictive analysis independently of the BI tool. The purpose of this article is to explain how to have these analyses on BI tools in addition to descriptive and diagnostic ones. More precisely, to prove the efficiency of BI tools in terms of data prediction, to explain the approach of integrating this type of analysis on these tools, to test different predictive models including machine learning and deep learning ones, and to qualify the precision of these different models. The application of these points will be done by working on a sample of data of a current important topic which is COVID-19.

10.2   Methodology

10.2.1   Dataset

For this article, we chose to work on COVID-19 data. The information was extracted from the database of the Center for Systems Science and Engineering (CSSE), Johns Hopkins University (JHU) [5]. The choice of this dataset was made according to several criteria:

- Use a reliable data source.
- Use a complete global dataset.
- Have diversified data (confirmed, deceased, and recovered cases).
- Have data broken down by several analysis axes, specifically the temporal one essential for the type of predictions we want to implement.
- Have data with a daily refresh to stay up to date.
- Have the ability to extract data in a format that can be easily interpreted by a BI tool.

10.2.2   BI Tool

The solution we chose to implement predictive analyses is Power BI: a collection of software services, applications, and connectors developed by Microsoft. With Power BI, we can connect to multiple different sources of data, combine them into a data model, and use this model to build visuals and dashboards. Power BI differs from other tools by its graphical richness, ease of use, and powerful features using advanced programming languages such as Python and R. For all these reasons and more, Power BI is positioned for the past few years as the leader of BI platforms, according to Gartner Magic Quadrant [6]. For this work, we used the free version: Power BI desktop.

10.2.3   Machine Learning Environment

Before starting the implementation of predictive analysis, it is useful to prepare a machine learning environment by making a set of installations: Anaconda, R and Python packages, specific libraries such as NumPy, Pandas, Matplotlib, Keras, Tensorflow, Sklearn, Statsmodels. These installations are done independently of Power BI, but once connected, the tool automatically detects their locations.

10.2.4   Data Integration

Once the installation was completed, we created a Power BI application and loaded the COVID-19 data using the following steps:

- Establish connections to data sources: In our case, the data is accessible through a URL, so we used the "Web" functionality of Power BI [7] to get the global statistics of confirmed, recovered, and deceased cases from the CSSE database.
- Prepare data: To facilitate the use of the collected data, some transformations have been done such as formatting dates, transposing data to have them in rows, and not in columns, and renaming fields.
- Create a data model: This means creating links between the three data sources (recovered, confirmed, and deceased cases) to be able to cross them on a single graph or to analyze them using the same analysis axis such as dates, countries, and so forth.

10.2.5   Visualization of Historical Data

Once the data model was established, we created some dashboards, starting with global distributions and then zooming in on Morocco as shown in Figures 10.1 and 10.2.

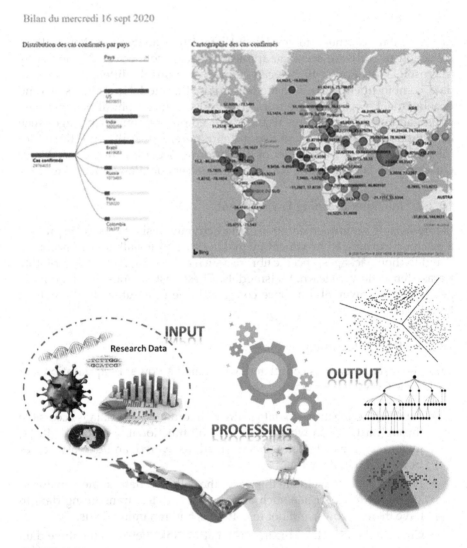

FIGURE 10.1
Worldwide distribution of confirmed cases.

At this stage, we have only restituted the historical data to get a clearer idea of the evolution of COVID-19. After analyzing the visuals, a question came to mind: What will be the situation in the coming days, especially in Morocco? To answer this question, it was necessary to do predictive analyses, and this is what we will explain in detail in the next chapter.

Bilan du mercredi 16 sept 2020

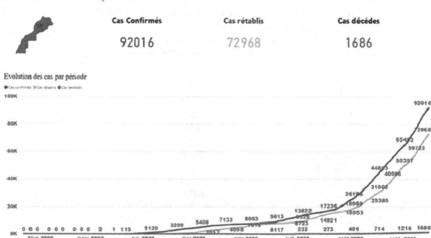

	Cas Confirmés	Cas rétablis	Cas décédes
	92016	72968	1686

FIGURE 10.2
Evolution of confirmed, recovered, and deceased cases in Morocco.

10.3 Results Interpretation

In this section, we will present the results of several predictive analyses realized in different ways on Power BI. For these analyses, we focused on the confirmed cases in Morocco, and we made predictions over 10 days from 16 September 2020. We will begin by presenting the results of each analysis. Then, we will qualify the exactitude of each predictive model to find the best one in terms of precision. Finally, we will explain the difference in the results of some models.

10.3.1 Results of Forecast Option

On some Power BI graphics, especially line charts, we can easily add a forecast to historical data by using the analytics pane [8]. Figure 10.3 and Table 10.1 show a view of the results returned after applying the necessary parameters.

10.3.2 Results of the Prediction Models Using R

In addition to the graphics available on Power BI, it is possible to import other ones with advanced functionalities from the Microsoft AppSource [9]. The graphics imported and used in this work are:

TABLE 10.1

Prediction Results of Confirmed Cases in Morocco Using the Forecast Option of Power BI

Date	Predicted confirmed cases
09/17/2020	93862
09/18/2020	95706
09/19/2020	97551
09/20/2020	99396
09/21/2020	101241
09/22/2020	103086
09/23/2020	104931
09/24/2020	106776
09/25/2020	108621
09/26/2020	110466

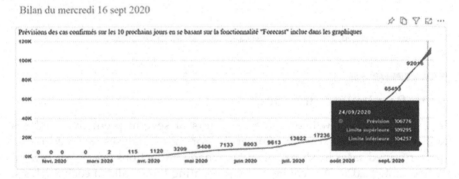

FIGURE 10.3

Prediction results of confirmed cases in Morocco using the forecast option of Power BI.

- Forecasting with ARIMA: This type of graphic applies one of the most commonly used methods for time-series forecasting, ARIMA (Auto-Regressive Integrated Moving Average).
- Forecast using Neural Network by MAQ Software: Neural networks are based on advanced data training and learning algorithms. They are generally recognized by their performance and for their ability to return results that are more or less close to reality.

To use these visuals, we have to import them and apply the necessary settings according to our analysis needs. Once the settings are applied, the graphics automatically run their appropriate algorithm on R and return the results as shown in Figures 10.4, 10.5 and Table 10.2.

TABLE 10.2

Prediction Results of Confirmed Cases in Morocco
Using the Models: ARIMA and ANN

Date	ARIMA	ANN
09/17/2020	92016	92762
09/18/2020	93895	93459
09/19/2020	95774	94108
09/20/2020	99532	94712
09/21/2020	101412	95274
09/22/2020	103291	95795
09/23/2020	105170	96278
09/24/2020	107049	96725
09/25/2020	108928	97139
09/26/2020	110807	97521

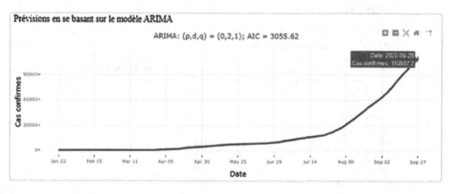

FIGURE 10.4

Prediction results of confirmed cases in Morocco using the ARIMA model.

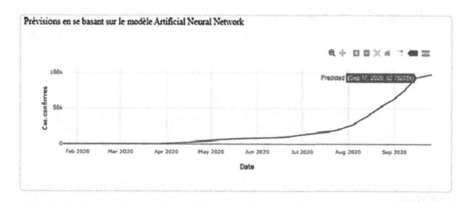

FIGURE 10.5

Prediction results of confirmed cases in Morocco using the ANN model.

10.3.3   Results of the Prediction Models Using Python

While analyzing the results returned by the ARIMA and ANN models using R, a question came to our mind: If we redo one of these models on Python, will we get the same results? To answer this question, we chose to reimplement the ARIMA model on Python by following the steps below:

- Realize the Python code of each model on the Jupyter Notebook and test the results referring to the tutorials [10] [11].
- Integrate the codes on Power BI [12] and link the Python prediction data to the existing one.
- Create new line charts on Power BI to visualize predicted data.

Figures 10.6 and Table 10.3 illustrate the results obtained.

TABLE 10.3

Prediction Results of Confirmed Cases in Morocco Using the ARIMA Model Implemented on Python

Date	Predicted confirmed cases using ARIMA Python
09/17/2020	93993
09/18/2020	95791
09/19/2020	97842
09/20/2020	99810
09/21/2020	101765
09/22/2020	103847
09/23/2020	105841
09/24/2020	107883
09/25/2020	109963
09/26/2020	112341

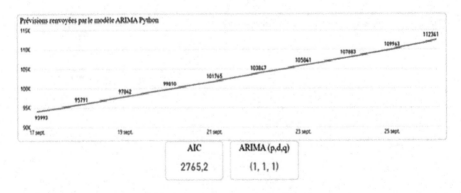

AIC	ARIMA (p,d,q)
2765,2	(1, 1, 1)

FIGURE 10.6
Prediction results of confirmed cases in Morocco using the ARIMA model implemented on Python.

TABLE 10.4

RMSE Results

Prediction method	RMSE
"Forecast" option	2812
ARIMA R	3073
ANN R	10551
ARIMA Python	1940

10.4 Qualification of the Predictive Models

Given the difference between the results obtained, it was necessary to qualify the precision of each model in order to identify the most reliable.

The evaluation of predictive models is done in several ways, such as calculating evaluation indices like the root mean square error (RMSE). The lower the RMSE value, the better the model is evaluated in terms of precision. The application of the RMSE requires actual and predicted data for the same period. So, we waited a few days to get the real statistics, then we used Excel to calculate the RMSE. Table 10.4 illustrates the results obtained.

The lowest RMSE corresponds to the ARIMA Python. We can then consider that the predictive analyses of this model are the most reliable in our context.

10.5 Conclusion

Through this article, we have proven that with the same BI tool we can have both historical and predictive analyses. By combining these two types of analyses, we were able to have an idea not only on the evolution of COVID-19 in Morocco in the previous months but also on the risks we can face in the coming days. It has also been shown that, with predictive models using artificial intelligence technologies, we can have better results in terms of data prediction.

What has been applied on the COVID-19 dataset is of course valid for any type of data that can be represented in a time series. Furthermore, the setting used for the days is also valid for the months and the years if we wish to extend the prediction area further.

Finally, it is important to talk about the added value of this work in terms of cost: Using a single BI tool for historical and future analysis costs less than using several tools, each one of them specialized in a particular type of analysis. Costs can also be better managed by applying optimization actions deduced from the predicted data.

References

[1] Louis Columbus (2020), What You Need to Know about BI in 2020. *Forbes Magazine*. Charts: "Functions Driving Business Intelligence", "Technologies and Initiatives Strategic to Business Intelligence Objectives by Industry", and "Technologies and Initiatives Strategic to Business Intelligence".

[2] Thierry Hamelin (2019), Du Big Data aux modèles prédictifs: où en sont les entreprises françaises.

[3] M.-L. Ivan and M. R. ˇaducu TRIFU (2016), "Using Business Intelligence Tools for Predictive Analytics in Healthcare System," IJACSA) *Int. J.Adv. Comput. Sci. Appl.*, vol. 7, no. 5, 2016.

[4] M. Rajarajeswari, Niranjana, S.V. (2019), "Customer Behavior Analysis and Prediction", vol. 21, no. 14, December 2019.

[5] JHU CSSE (2019), COVID 19/csse_covid_19_data/csse_covid_19_time_series at master · CSSEGISandData/COVID-19 · GitHub.

[6] Gartner Magic Quadrant for Analytics and Business Intelligence Platforms (2020), 2020 Gartner Magic Quadrant | Power BI (microsoft.com).

[7] Microsoft Documentation (2019), Connect to a webpage from Power BI Desktop - Power BI | Microsoft Docs.

[8] Microsoft Documentation (2020), Use the Analytics Pane in Power BI Desktop – Power BI | Microsoft Docs.

[9] Microsoft AppSource (2020), Microsoft AppSource – l'emplacement pour les applications métier.

[10] Thomas Vincent (2017), A Guide to Time Series Forecasting with ARIMA in Python 3, ARIMA Time Series Data Forecasting and Visualization in Python | DigitalOcean.

[11] Ian Felton (2019), A Quick Example of Time-Series Prediction Using Long Short-Term Memory (LSTM) Networks, The Startup | Medium.

[12] Microsoft Documentation (2020), Run Python Scripts in Power BI Desktop – Power BI | Microsoft Docs.

[13] Boris Efraty (2018), Powerbi-visuals-forcastingarima/script.r at master · microsoft/powerbi-visuals-forcastingarima · GitHub.

Index

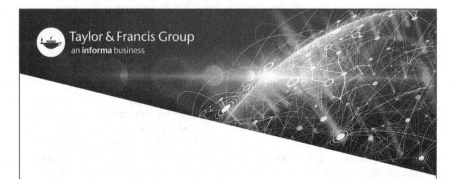

Printed in the United States
by Baker & Taylor Publisher Services